SpringerBriefs in Cancer Research

W0235092

For further volumes:
http://www.springer.com/series/10786

Kristy A. Brown · Evan R. Simpson

Obesity and Breast Cancer

The Role of Dysregulated Estrogen Metabolism

 Springer

Kristy A. Brown
Evan R. Simpson
Metabolism and Cancer Laboratory
Prince Henry's Institute
Clayton, VIC
Australia

ISBN 978-1-4899-8001-4   ISBN 978-1-4899-8002-1   (eBook)
DOI 10.1007/978-1-4899-8002-1
Springer New York Heidelberg Dordrecht London

Library of Congress Control Number: 2013954031

Printed on acid-free paper

Springer is part of Springer Science+Business Media (www.springer.com)

Acknowledgments

This work was supported by NHMRC (Australia) Project Grant # GNT1005735 to KAB and ERS, the Victorian Government, through the Victorian Cancer Agency funding of the Victorian Breast Cancer Research Consortium to ERS and KAB, and by the Victorian Government Operational Infrastructure Support Program. KAB is supported by an NHMRC (Australia) Career Development Award GNT1007714. ERS is supported by an NHMRC (Australia) Senior Principal Research Fellowship GNT0550900.

Contents

Introduction

Despite the decreased mortality rates attributable to breast cancer in women, the number of cases diagnosed has steadily increased over the past 30 years. This is believed to be due, at least in part, to an increased prevalence of obesity not only in the Western world, but also in other parts of the world where obesity has only recently reached epidemic proportions. The risk of breast cancer increases with age and a strong correlation between obesity and the risk of breast cancer in postmenopausal women is well established. The majority of postmenopausal breast cancers are hormone receptor positive and rely heavily on estrogens produced from the adipose tissue for growth. The enzyme responsible for the final and key step in estrogen biosynthesis, aromatase, is increased in the adipose tissue in response to factors produced in obesity, including adipokines, inflammatory cytokines, and prostaglandins, as well as insulin. Novel therapies are now being considered in light of evidence suggesting that obesity may affect current endocrine therapy, as well as the identification of novel pathways involved in estrogen regulation, including metabolic pathways that can be targeted by drugs currently used for the treatment of other obesity-related diseases. The current work aims to provide a comprehensive view of the relationship between obesity and breast cancer with particular emphasis on the role of dysregulated estrogen metabolism.

Chapter 1
Estrogens, Adiposity and the Menopause

A woman's reproductive years are characterized by finely tuned waves of circulating sex hormones which are dictated by her menstrual cycle and in turn, influence many, if not all, of her physiological functions. During the menopausal transition, ovarian steroid biosynthesis ceases leading to whole body effects including bone loss, a change in fat deposition and weight gain. These effects are believed to be largely due to deficiencies in circulating estrogens, although androgens are also suspected of being involved. Evidence to support this hypothesis has come from studies using animal models. Estrogen receptor α (ERα) knockout female mice have increased fat pad weights, adipocyte size and number [1]. Similarly, aromatase knockout mice, where estrogen biosynthesis is abolished, progressively accumulate intra-abdominal fat compared to wild type littermates with increased adipocyte volume [2]. As proof-of-principle, replacement of estrogens in these animals reduces omental and infrarenal adipose tissue weights [3]. Both animal models also display impaired glucose tolerance and insulin resistance. Studies in women have demonstrated that the menopausal transition, where estrogen levels decrease rapidly and androgen levels remain steady, is associated with the accumulation of intra-abdominal fat (Fig. 1.1) [4], believed to be due to a decrease in resting metabolic rate and physical activity [5]. Estrogens act to regulate adiposity both centrally and within the periphery. Knockout of ERα in the central nervous system, specifically in the ventromedial nucleus, leads obesity and the metabolic syndrome due to changes in energy expenditure [6]. Estrogens have also been shown to inhibit food intake and loss of gonadal steroids is associated with hyperphagia and weight gain [7]. In the periphery, estrogens play an important role in muscle and adipose tissue metabolism. Adipocytes express ERα [8] and estrogens have been shown to reduce adipocyte volume by inhibiting the expression of genes involved in fatty acid uptake and lipogenesis [9–12]. In the muscle, estrogens stimulate the oxidation

K. A. Brown and E. R. Simpson, *Obesity and Breast Cancer*,
SpringerBriefs in Cancer Research, DOI: 10.1007/978-1-4899-8002-1_1,
© The Author(s) 2014

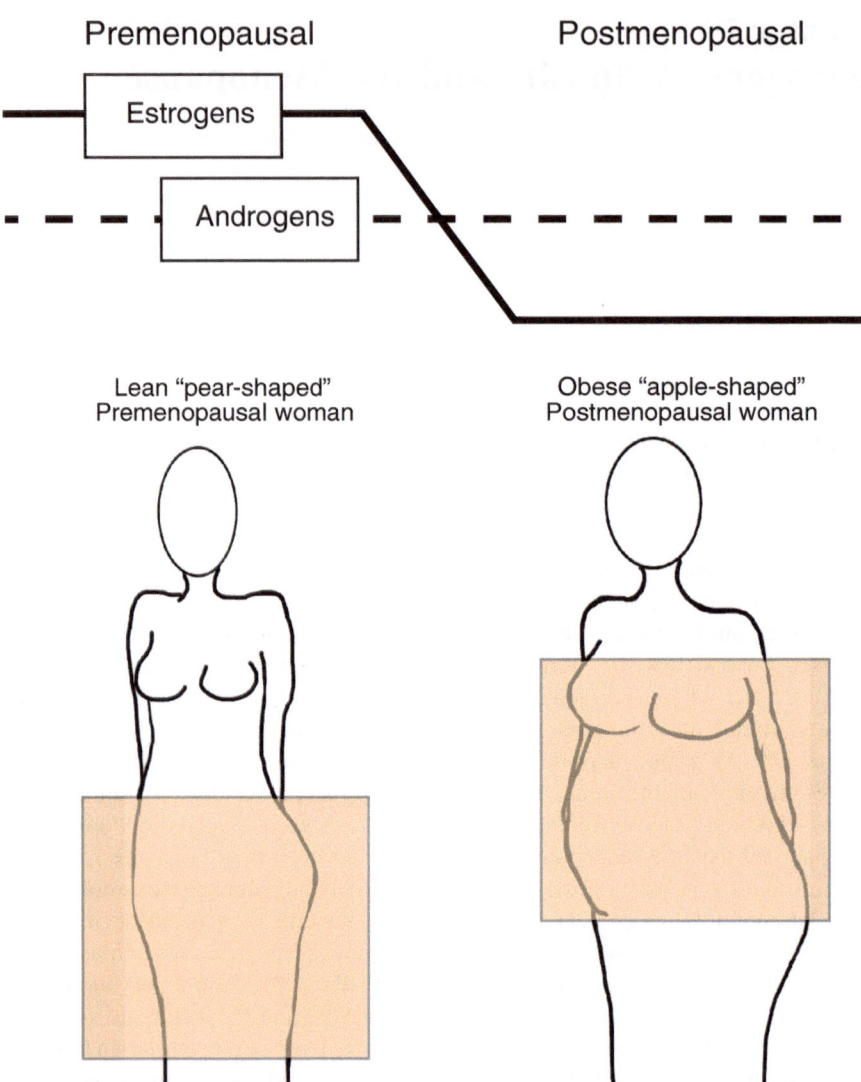

Fig. 1.1 Change in body shape during the menopausal transition. A decrease in estrogens after menopause, coupled to an increased ratio of androgens to estrogens, has been suggested to impact fat accumulation and accumulation of fat at specific depots (*shaded*). Namely, premenopausal women tend to accumulate fat subcutaneously around the buttocks and thighs whereas postmenopausal women tend to accumulate fat intra-abdominally

of fatty acids and lipolysis [13]. Interestingly, the loss of ovarian steroids has also been associated with an increase in adipose tissue inflammation [14], a phenomenon that is reversed in the presence of estrogens [15].

Chapter 2
The Link Between Obesity and Breast Cancer Risk: Epidemiological Evidence

2.1 BMI and Breast Cancer Risk

BMI is routinely used to qualify an individual's adiposity, yet it is simply a measure of an individual's mass (kg) divided by their height^2 (m^2). According to the WHO international classification, individuals with a BMI between 18.5 and 24.99 are considered healthy, whereas those with a BMI between 25 and 29.99 or of 30 and above are considered overweight or obese, respectively. Recently and due to the growing number of individuals with BMI values above 30, it has also become necessary to further subdivide the obese category into three classes; obese class I (BMI 30–34.99), obese class II (BMI 35–39.99) and obese class III (BMI $\geq$ 40) [16].

Obesity rates have doubled since 1980 and in 2008, were estimated at 300 million for adult women [17]. A BMI above 25 increases the risk of a number of diseases, including heart disease and stroke, diabetes, musculoskeletal disorders, as well as cancers of the endometrium, colon and breast. An exponential increase in the number of publications examining the association between BMI and breast cancer has occurred over the last two decades (Fig. 2.1). As of July 1st, 2013, using the search terms "body mass index" and "breast cancer" in Pubmed returned 2221 publications, 232 were published in 2012 alone. This highlights the burden of these co-morbidities as well as the advances made in recent years, in particular with regards to understanding the epidemiological link and effect of obesity on breast cancer management.

Obesity is associated with an increased risk of breast cancer, and is also positively associated with tumor size and a higher probability of having positive axillary lymph nodes and faster growing tumors [18–20]. Interestingly, higher energy intake also increases the risk of breast cancer [21]. It is well accepted that obesity increases the risk of developing breast cancer after menopause, and it has even been suggested that up to 50 % of postmenopausal breast cancers are attributable to obesity [22]. However, the degree of increased relative risk and whether or not this also holds true for premenopausal women is contentious. A number of meta-analyses have been performed in recent years examining the

K. A. Brown and E. R. Simpson, *Obesity and Breast Cancer*,
SpringerBriefs in Cancer Research, DOI: 10.1007/978-1-4899-8002-1_2,
© The Author(s) 2014

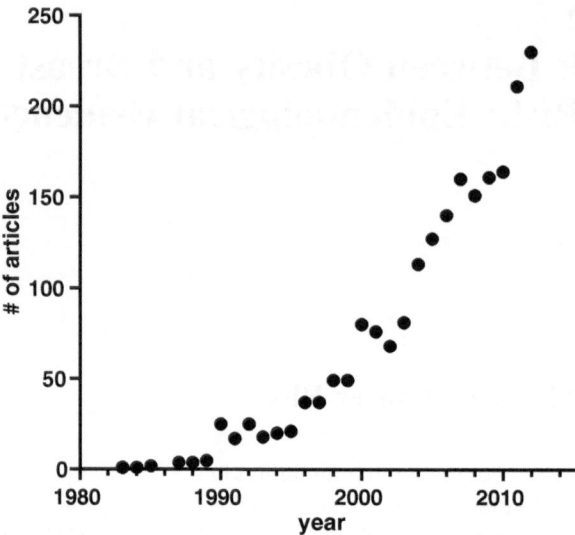

Fig. 2.1 Number of publications on BMI and breast cancer over the last 3 decades. A Pubmed search was performed using search terms "body mass index" and "breast cancer" and plotted as number of articles per year

Table 2.1 Effect of high BMI on breast cancer risk with respect to menopausal status

Type of study	Menopausal status	RR (95 % CI)	Reference
Meta-analysis	Premenopausal	0.93 (0.86–1.02)	Cheraghi et al. [23]
	Postmenopausal	1.15 (1.07–1.24)	
Meta-analysis	Premenopausal	0.98 (0.97–0.99)[a]	Bergström et al. [24]
	Postmenopausal	1.02 (1.02–1.03)[a]	
Meta-analysis	Premenopausal	1.43 (1.23–1.65)	Pierobon et al. [29]
(Triple negative)	Postmenopausal	0.99 (0.79–1.24)	
Meta-analysis	Postmenopausal	1.19 (1.05–1.34)	Key et al. [31]

[a] per unit increase in BMI

effect of BMI on breast cancer with age (Table 2.1). In 2012, Cheraghi et al. performed a meta-analysis of 50 studies, 15 cohort studies and 35 case-control studies involving 2,104,203 and 71,216 participants, respectively [23]. There was no significant effect of BMI on breast cancer risk in premenopausal women, but a direct and significant correlation was observed between BMI and breast cancer risk in postmenopausal women. This was consistent with findings from Bergstrom et al. who demonstrated that a one unit increase in BMI was associated with a 2 % increased risk of developing breast cancer in postmenopausal women [24], a relationship that was not found in the premenopausal group. A number of studies have also described obesity as strongly protective against breast cancer in premenopausal women [25–28]. This has been attributed, at least in part, to a greater number of anovulatory menstrual cycles and hence, decreased lifetime exposure to estrogens. Nevertheless, a study Biglia et al. demonstrated that high BMI was significantly associated with larger sized tumors in both pre- and postmenopausal women [20]. In this case, obese premenopausal women displayed more vascular infiltration and metastasis to axillary lymph nodes compared to healthy weight women. Moreover, a meta-analysis by Pierobon et al. revealed that obesity is a significant risk factor for triple negative breast cancers in pre- but not postmenopausal women [29]. Interestingly, there is also evidence for an association of BRCA1 mutations and BMI with breast cancer risk in premenopausal women (P = 0.045) [30].

2.2 BMI and Breast Tumor Hormone Receptor Status

The types of tumors which occur in obese pre- and postmenopausal women have also been examined. In Japanese women, and consistent with the study of Pierobon et al., obesity is associated with an increased risk of triple negative tumors prior to menopause [32]. Conversely, the majority of luminal B breast cancers tended to occur in obese postmenopausal women. In the study by Biglia et al., BMI was not associated with tumor type in premenopausal women, however, there was a significant association between BMI and estrogen receptor (ER)/progesterone receptor (PgR)-positive tumors in postmenopausal women [20]. A recent case-control study by John et al. demonstrated that weight gain of ≥ 30 kg between early adulthood and menopause was associated with a 1.53-fold increased risk of developing hormone receptor-positive breast cancer amongst all women studied, while non-hispanic white women were 3.82-fold more likely to develop breast cancer compared to women who's weight remained stable [33]. A similar study was undertaken by Krishnan et al. who demonstrated that while weight at 18–21 years was not associated with risk of breast cancer after menopause, an increase in weight during adulthood was positively associated with the increased risk of PgR-positive breast cancers after menopause (HR per 5 kg/m^2 gain in BMI: 1.43; 95 % CI: 1.23–1.66) [34]. Taken together, these studies demonstrate that obesity-related postmenopausal tumors are largely dependent on steroid hormones for growth. Conversely, obese premenopausal women tend to develop triple negative tumors, suggesting that other obesity-associated factors may play pivotal roles in tumor development.

2.3 Obesity and Mammographic Density

Mammographic density is one of the strongest predictors of breast cancer risk and reflects the relationship between dense epithelial and non-epithelial cell abundance, as well as acellular components including collagen [reviewed in 35]. The relationship between obesity, mammographic density and breast cancer, however, is still unresolved. This is largely due to the fact that obese women tend to have less dense breasts, as measured by percentage breast volume and absolute dense breast volume [36]. There are some key findings, however, that suggest that the relationship is more complex [37]. There is considerable heterogeneity of dense and non-dense areas within the breast and this reflects important differences in tissue composition, including the presence of estrogen-producing stromal cells. Indeed, aromatase expression [38] and the ratio of parent estrogen compounds (estrone and estradiol) to estrogen metabolites [39] are higher in dense areas of the breast compared to non-dense areas. Therefore, additional studies examining the differences in these areas are warranted in order to elucidate whether a relationship between obesity, mammographic density and breast cancer risk exists.

2.4 Waist-to-Hip Ratio and Breast Cancer

The often reported inverse association between BMI and breast cancer risk in premenopausal women has caused much controversy. This is largely due to the fact that BMI reflects overall adiposity rather than specific sites of adipose depots. More recently, waist-to-hip ratio has gained popularity as a measure of unhealthy weight gain and a study by Amadou et al. demonstrated that each 0.1 unit increase in waist-to-hip ratio was associated with an increased relative risk of 1.19 (95 % CI: 1.15–1.24) of premenopausal breast cancer irrespective of ethnicity [40]. Additional studies, however, are required in order to determine whether waist-to-hip ratio should be used in assessing a premenopausal woman's risk of breast cancer.

2.5 The Metabolic Syndrome, Diabetes Mellitus and Breast Cancer

Overweight and obesity significantly increases the risk of developing type 2 diabetes mellitus (T2DM). Namely, an overweight individual carries a threefold increased risk of T2DM whereas obese individuals are seven times more likely to develop T2DM [41]. With increased obesity rates has come an increase in the prevalence of T2DM. It is now estimated that approximately 7 % of Americans have T2DM as a consequence of the development of insulin resistance (reviewed in [42]). This figure is not only characteristic of US populations but also represents a growing trend in other developed and developing countries. The risk of death in individuals with T2DM is twofold [43] and occurs as a result of a number of diabetes-related complications including heart disease and stroke, as well as infectious diseases, degenerative disorders and several types of cancers [44].

Several studies have examined the association between diabetes and breast cancer risk and a meta-analysis was recently performed [45]. From observational studies, the summary relative risk of developing breast cancer in women with T2DM compared to those without was 1.17 (95 % CI: 1.13–1.63), whereas prospective and retrospective studies had a summary relative risk of 1.23 (95 % CI: 1.12–1.35) and 1.36 (95 % CI: 1.13–1.63), respectively. Of interest, studies that adjusted for BMI had a lower summary relative risk than those that didn't (1.16 vs. 1.33, respectively). This suggests that BMI itself is a risk factor for breast cancer, but the remaining increased risk also supports a role for diabetes independent of BMI. Indeed, a study of women in Eastern China demonstrated that women with a history of diabetes were 3.5 times more likely to develop breast cancer than women who didn't (odds ratio: 3.556; 95 % CI: 0.904–13.994), whereas having a high BMI index was associated with a 1.5-fold increased risk of developing the disease (odds ratio: 1.528; 95 % CI: 1.083–2.155) [46].

The relationship between the metabolic syndrome and breast cancer risk has also been examined in a recent meta-analysis [47]. Nine studies were included in the meta-analysis and overall metabolic syndrome was shown to be associated with a 52 % increase in breast cancer risk. This study also examined associations between BMI, hyperglycemia, blood pressure, triglycerides and cholesterol in relation to breast cancer risk.

2.6 Breast Size and Breast Cancer

Few studies have examined whether or not an association exists between breast size and breast cancer risk. In 2006, a prospective study examining breast size and premenopausal breast cancer incidence demonstrated that healthy weight women with a bra cup size of "D or larger" had a significantly higher incidence of breast cancer than women who reported "A or smaller" [48]. The association was lost in women with a higher BMI. In a study by Markkula et al., a prospective breast cancer cohort study (n = 772) examined the characteristics of women with breast cancer who had a larger breast size [49]. Findings demonstrate that breast that were larger than 850 ml in volume tended to have larger tumor size, more advanced histological grade and more axillary node involvement. Much debate relating to whether increased risk of breast cancer in larger breasted women is in fact due to most women with larger breasts having a higher BMI. Nevertheless, after adjusting for BMI, this study demonstrated that in patients with ER-positive tumors, breast size was an independent predictor of disease-free and distant metastasis-free survival.

2.7 Effect of Obesity on Disease-Free Survival

A number of studies have examined the impact of BMI on breast cancer recurrence and death. A retrospective cohort study by Kamineni et al. demonstrated that obese women with early-stage breast cancer had a significantly increased risk of recurrence (HR 2.42; 95 % CI: 1.34–4.41) and breast cancer-related death (HR 2.41; 95 % CI: 1.00–5.81) within 10 years of diagnosis compared to healthy-weight women [19].

Druesne-Pecollo et al. performed a meta-analysis of clinical studies whereby they examined the impact of excess body weight on second primary cancer risk after breast cancer across thirteen prospective, five cohort and eight nested case-control studies [50]. Findings demonstrate that obesity increases the relative risk of developing breast, contralateral breast, endometrial and colon second primary cancers. Elevated serum total cholesterol, triglycerides, low-density lipoprotein cholesterol and the ratio between low-density and high density lipoprotein cholesterol, known to occur in obesity and the metabolic syndrome, have also been shown to be associated with a significantly higher distant metastasis rate [51]. A study by Forsythe et al. demonstrated that breast cancer survivors who were overweight or obese also had higher pain compared to healthy weight women and when examined longitudinally, weight gain above 5 % was positively associated with above-average pain [52]. Finally, diabetes has also been shown to be positively associated with risk of breast cancer-associated death after controlling for BMI (relative risk: 1.16; 95 % CI: 1.03–1.29) [53].

Chapter 3
Adipose-Derived and Obesity-Related Factors and Breast Cancer

3.1 Adipokines

It is now clear that the adipose is an important endocrine organ. Considering the vast evidence demonstrating a link between obesity and breast cancer, it is not surprising that attention has turned to the role of factors produced by the adipose, termed adipokines, as major drivers of breast cancer growth via direct effects on cell proliferation and indirect effects on estrogen biosynthesis (Fig. 3.1). These adipokines include hormones, growth factors and cytokines and recent evidence suggests that over 250 different adipokines are secreted by adipocytes [54]. The most widely studied adipokines are leptin and adiponectin, which are peptide hormones that are differentially regulated depending on whole body energy states.

Adiponectin is produced by healthy mature adipocytes and its levels are inversely associated with obesity [55]. Many reports have suggested that adiponectin is protective against breast cancer development and progression [reviewed in 56]. In a study by Gulcelik et al., serum adiponectin levels were evaluated in 87 breast cancer patients and compared to cancer-free women [57]. Circulating adiponectin levels were found to be significantly lower in affected individuals compared to healthy controls (8,583 ± 2,095 ng/ml vs. 13,905 ± 3,263). The anti-cancer effects of adiponectin are hypothesized to be mainly due to the effect of adiponectin to reduce inflammation and increase insulin sensitivity [58]. However, effects of adiponectin on hormone biosynthesis and direct effects of adiponectin on cancer growth have also been described.

Conversely, leptin levels are higher in obese individuals, and higher leptin levels are significantly associated with an increase in breast cancer risk [59, 60]. Moreover, a prospective observational study by Macciò et al. demonstrated that leptin was also an independent predictor of tumor classification and TNM stage in postmenopausal women [61]. Leptin receptor expression is common in breast cancer [62], suggesting that leptin can act directly on tumor cells to modulate tumor growth. Leptin has been shown to activate a number of mitogenic pathways in cancer cells, including phosphoinositide 3-kinase/protein kinase B (PI3 K/AKT), mitogen-activated protein kinase (MAPK), mammalian target of rapamycin

K. A. Brown and E. R. Simpson, *Obesity and Breast Cancer*,
SpringerBriefs in Cancer Research, DOI: 10.1007/978-1-4899-8002-1_3,
© The Author(s) 2014

(mTOR) and signal transducer and activator of transcription 3 (STAT3) [63, 64]. In vitro, leptin has been shown to increase heat shock protein 70 (HSP70) expression in MCF-7 breast cancer cells [65], which in turn is known to increase cell proliferation [66]. In mouse models, leptin was shown to be responsible for the increase in expression of cyclin D1, a cell-cycle control protein necessary for mammary gland development, in MMTV-Wnt1 transgenic mice [67].

A less well characterized adipokine, nicotinamide phosphoribosyl-transferase (Nampt), also known as visfatin, has also been implicated in cancer [reviewed in 68]. This pleiotropic hormone has been shown to be secreted from visceral adipose depots and acts to regulate a number of metabolic processes including NAD biosynthesis. It has been shown to promote cell proliferation, inflammation and angiogenesis, and inhibit apoptosis. Nampt has also been shown to be secreted from tumor cells where it can act in an autocrine manner. Similarly, resistin, involved in stimulating low density lipoprotein production from the liver, has been shown to be positively associated with tumor size and stage, as well as ER status [69].

Breast levels of the adipokines leptin and adiponectin, and their relationship to blood levels have also been examined [70]. There was a strong positive correlation between blood and breast leptin in healthy weight women, whereas the strongest association between blood and breast adiponectin was seen in obese women. Interestingly, leptin levels increased more substantially with increasing BMI in the breast than in plasma, suggesting that small increases in weight may be associated with more important pro-proliferative effects in the breast.

3.2 Inflammatory Factors

Obesity is a recognized state of low grade chronic inflammation and inflammatory factors, including cytokines and prostaglandins, have been implicated in the development and progression of breast cancer, again via direct and indirect mechanisms (Fig. 3.1). Cytokines produced in obesity include interleukin (IL)-1β, IL-6, IL-8, tumor necrosis factor α (TNFα), and a number of chemokines. This causes the increased recruitment of immune cells which then further drives inflammation.

Inflammation has also been shown to inhibit adipocyte differentiation and inflammatory factors stimulate leptin secretion from preadipocytes [71]. Moreover, leptin stimulates macrophage maturation further enhancing the pool of local inflammatory and angiogenic factors, including TNFα, fibroblast growth factor (FGF), epidermal growth factor (EGF), VEGF, IL-6 and IL-8 within the obese adipose [72, 73]. High levels of circulating inflammatory factors are associated with a worse prognosis in women who develop breast cancer.

In addition to systemic changes in inflammatory factors, obesity is also associated with changes within the breast. Recent studies have demonstrated that macrophage infiltration into the obese breast leads to the formation of crown-like structures associated with an increase in prostaglandin E$_2$ (PGE$_2$) and nuclear

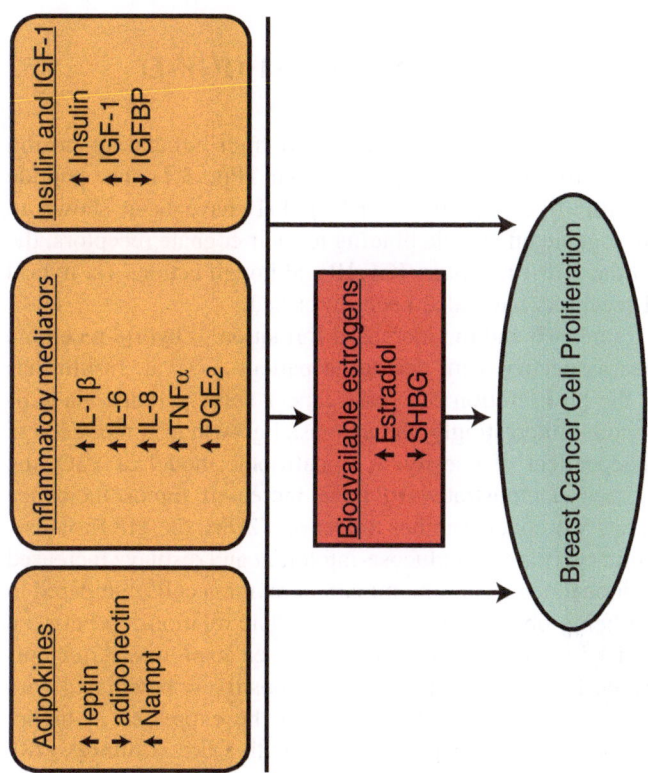

Fig. 3.1 Direct and indirect effects of adipokines, inflammatory mediators and insulin and IGF-1 on tumor cell growth. In obesity, hyperinsulinemia and high levels of certain adipokines, IGF-1 and inflammatory mediators can act directly on tumor cells to stimulate proliferation. These factors also stimulate estrogen production from the adipose which in turn further stimulate tumor cell proliferation

factor κB (NF-κB) [74, 75]. Inflammation also occurs at the site of tumor growth and inflammatory mediator expression at the site of primary breast tumors is associated with poor patient outcome. Possible mechanisms explaining the pro-proliferative effects of these inflammatory mediators have been proposed. Similar to leptin, cytokines act to increase cancer cell proliferation via the activation of a number of mitogenic signaling pathways, including PI3 K, MAPK, mTOR and STAT3 [76]. These inflammatory mediators also stimulate tumor growth by inducing estrogen biosynthesis from preadipocytes or adipose stromal cells and this relationship will be explored in more detail in Chap. 4.

3.3 Insulin and Insulin-Like Growth Factor-1 (IGF-1)

Obesity is often associated with hyperinsulinemia and high circulating insulin levels are believed to play a role in driving tumor growth (Fig. 3.1). The peptide hormones insulin and insulin-like growth factor-I (IGF-I) have been shown to mediate metabolic and mitogenic effects via binding to their cognate receptors, the insulin receptor (IR) and the IGF-I receptor (IGF-IR), although actions via hybrid receptors (insulin/IGF-I receptor) have also been described.

Changes in receptor abundance and the increased formation of hybrid receptors on tumor cells have been associated with decreased patient survival. Insulin has been shown to increase the proliferation of breast cancer cells, perhaps via activation of the hybrid receptors. Interestingly, the increased growth of tumor cells in T2DM appears to be independent of obesity. A lipoatrophic model of T2D, the A-ZIP/F-1 mouse, has been demonstrated to have increased tumor incidence despite having no white fat. Moreover, another model of T2DM, the MKR mouse, is hyperinsulinemic, insulin-resistant and glucose-intolerant and displays increased growth of orthotopically inoculated mouse mammary carcinoma cells compared to control mice, despite not being obese [reviewed in 77]. The relationship between fasting insulin levels and breast cancer risk has also been explored. After correcting for obesity, elevated fasting insulin levels were positively associated with breast cancer risk. Interestingly, IGF-I is also found to be expressed at higher levels in ER-positive tumors when compared to ER-negative breast tumors [78]. Moreover, higher IGF-1 has been shown to be associated with higher percent mammographic density in postmenopausal women with a healthy BMI, an association not seen in premenopausal women or women with a higher BMI [79].

3.4 Estrogens

Considering that the majority of obesity-related postmenopausal breast cancers are estrogen-dependent, it is not surprising that considerable effort has been devoted to understanding the regulation of this postmenopausally adipose-derived steroid hormone. Consequently, Chap. 4 is devoted to a comprehensive overview of the regulation of estrogens in adipose tissue and of the enzyme responsible for the key and final step in estrogen biosynthesis, aromatase.

Chapter 4
Estrogen Biosynthesis

4.1 Source of Estrogens in Pre-versus Postmenopausal Women

In the premenopausal woman, the main site of estrogen production is the ovaries, and with every menstrual cycle, rising follicle stimulating hormone levels induce estrogen biosynthesis from granulosa cells of the developing follicles [80]. Estrogens produced from the ovaries act in an endocrine fashion to modulate gonadotrophin secretion, growth of the uterine lining, as well as maintain normal function of a wide range of tissues, including the breast, bone and brain (Fig. 4.1). In pregnancy, the main site of estrogen production becomes the placenta, and more precisely in the syncytiotrophoblast [81]. After menopause, however, estrogens are not found at high levels in circulation and until recently were barely detectable in human plasma. Nevertheless, estrogens continue to play a vital role in maintaining organ function after menopause and this is possible via the local production of estrogens that then act in an autocrine and paracrine manner (Fig. 4.1). The most important source of estrogens in these women is also one of the largest endocrine organs, the adipose, and this production of estrogens increases as a function of obesity and aging [82–84]. Interestingly, breast tissue estrogen levels in postmenopausal women are similar to those found in premenopausal women, despite 10–15-fold higher levels of circulating estrogens in premenopausal women [85]. Of importance, breast adipose tissue also has the capacity to aromatize androgens and is the basis for much of the current work aimed at devising breast-specific aromatase inhibitors that will inhibit breast cancer growth without affecting the bone and brain, where estrogens have beneficial effects [86].

4.2 The Aromatase Enzyme

The aromatization of C19-androgens into C18-estrogens is the key and final step in estrogen biosynthesis and this reaction is uniquely catalyzed by the aromatase enzyme. Aromatase is a member of the cytochrome P450 superfamily of

K. A. Brown and E. R. Simpson, *Obesity and Breast Cancer*,
SpringerBriefs in Cancer Research, DOI: 10.1007/978-1-4899-8002-1_4,
© The Author(s) 2014

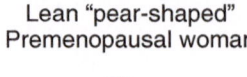

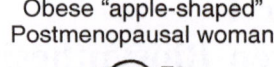

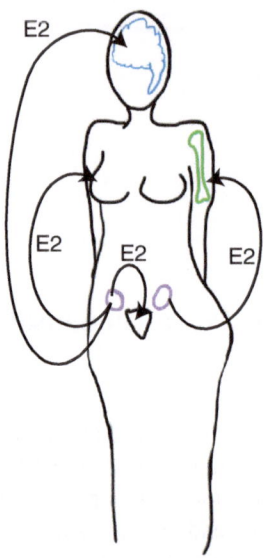

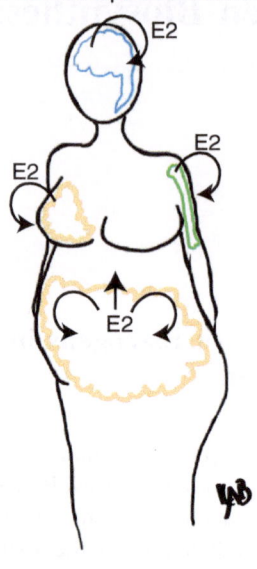

Fig. 4.1 Source of estrogens in lean premenopausal and obese postmenopausal women. In premenopausal women, the main source of estrogens is the ovaries. Estrogens then act in an endocrine fashion to regulate function of tissues including the brain (*blue*), bone (*green*) and uterus (*black*). After menopause, extragonadal tissues produce low levels of estrogens which then act in an autocrine and paracrine fashion. In obese postmenopausal women, the adipose becomes the predominant site of estrogen biosynthesis contributing to an increase in the risk of developing estrogen-dependent breast tumors

hemeproteins which contains over 6,000 members. Aromatase belongs to gene family CYP19 and is the family member CYP19A1. Most members of this superfamily have molecular weights in the range of 50 kDa, and in keeping with this, aromatase has 503 amino acids. The sequence has regions of marked homology with other members of the superfamily, notably a putative N-terminal membrane-spanning region, I-helix, Ozols and heme-binding region. These enzymes require molecular oxygen for their catalytic activity as well as a source of reducing equivalents. In the case of the mammalian members, these are derived from NADPH. The NADPH interacts with and passes electrons to the flavoprotein NADPH-cytochrome P450 reductase which in turn transfers the electrons to the aromatase cytochrome P450. These proteins are located at the endoplasmic reticulum [87]. The aromatase protein, expressed from a full-length cDNA insert, was shown to catalyse the aromatization of androstenedione and testosterone, and in the case of the human placenta, 16α-hydroxyandrostenedione. The reaction was also shown to be inhibited by known aromatase inhibitors. Conversion of C19-steroids into C18-estrogens occurs in a complex three-step process catalysed by this single polypeptide chain and requires three moles of cofactor NADPH and three moles of oxygen for every mole of C19 androgen converted (Fig. 4.2). Light has recently been shed on the precise molecular symphony that gives rise to the formation of estrogens as a consequence of the elucidation of the crystal structure of aromatase by the group of Ghosh and colleagues in Buffalo, NY [88]. The proposed reaction mechanism highlights the importance of residues such as D309 in the catalytic reaction mechanism as well as that of C437. This cysteine is present in the heme-binding region and occupies the 5th coordination position of the heme iron, and is uniquely common to all P450 enzymes. Specific to aromatase, in the third step of the reaction sequence, the A ring aromatization step, D309 is shown to be involved in enolization of the 3-oxo group, and removal of the 2β-hydrogen is facilitated by the carbonyl group of A306. The reaction mechanism of aromatase is very specific essentially limited to C19 steroids due to a highly specific binding cleft at the catalytic site.

4.3 Local Aromatase Expression in Breast Cancer

The source of estrogens driving tumor growth in postmenopausal women remains an area of contention. Two schools of thought exist. On the one hand, it is believed that estrogens produced locally within the breast serve to stimulate breast cancer cell growth whereas the second contends that circulating estrogens are taken up by the breast and drive tumor growth. Both arguments have merit and the strength of each argument depends on adipose-derived estrogens as a driver of tumor cell growth.

Evidence to support circulating estrogens as drivers of tumor growth are largely epidemiological and rely on findings demonstrating that postmenopausal breast cancer incidence is positively correlated with body fat content and serum estrogen

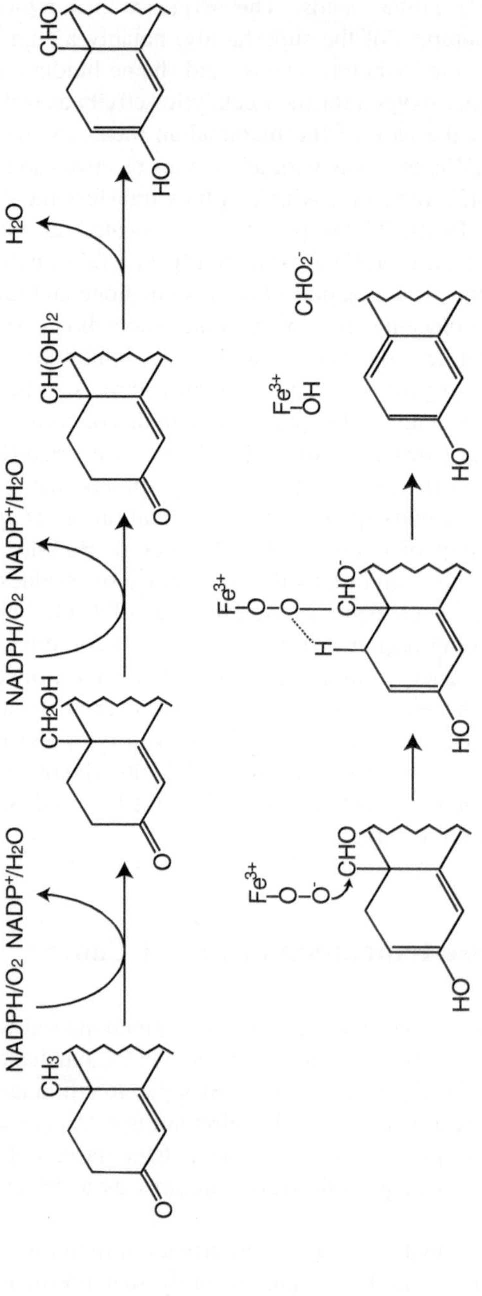

Fig. 4.2 Mechanism of aromatization of estrogens into androgens by the aromatase enzyme

levels [89, 90]. This argument is also supported by findings demonstrating that intratumoral estrogens are positively associated with tumor ERα expression, suggesting that uptake of estrogens contributes to tumor growth [91]. These two studies are not inconsistent, however, with circulating estrogens arising as a consequence of adipose tissue estrogen biosynthesis or originating from the adjacent breast adipose.

There is substantial evidence to support local breast estrogen production as a major driver of breast tumor growth. Of note, O'Neill et al. examined mastectomy tissue from breast cancer surgery and demonstrated that aromatase activity is highest in the breast quadrant which contains the tumor [92]. These findings were later supported by studies examining aromatase transcript, protein and activity in similar tissue [93–97]. This would suggest one of two scenarios, either that tumors tend to originate in areas which have high estrogen biosynthesis or that factors produced by tumors stimulate aromatase expression and tumor growth. Reports to date would suggest that both scenarios exist. As support for the hypothesis that breast tumors originate in a region of the breast with high estrogen levels, a recent study has demonstrated that mammographically dense breasts have high aromatase expression [38]. Using immunohistochemistry performed on breast core biopsies, the authors demonstrate that areas from dense areas of the breast had the highest immunoreactivity for aromatase with highest levels being seen in the stroma compared to other cell types and non-dense areas of the breast. The hypothesis that this elevated aromatase expression may lead to an increased risk of breast cancer is not only supported by epidemiological data demonstrating a clear link between mammographic density and breast cancer incidence, but also in animal models whereby overexpression of aromatase in the mammary gland of ovariectomized mice leads to breast hyperplasia, a phenomenon that is reversible in the presence of the aromatase inhibitor letrozole [98].

The relative contribution of estrogen biosynthesis within various cell types within the breast has also been explored. Adipose stromal cells express aromatase and have measurable aromatase activity [99], and much of the work pertaining to aromatase regulation within the breast has been performed using adipose stromal cells in monolayer culture. Nevertheless, aromatase immunoreactivity has been detected in both the adipose stroma and tumorous epithelium of the breast [100–102] and factors that contribute to the regulation of aromatase expression in both cell types is discussed below.

4.4   The *CYP19A1* Gene and Tissue-Specific Expression

Unlike other steroidogenic enzymes, including 17β-hydroxysteroid dehydrogen-ases, which undergo tissue-specific regulation due to being encoded by different genes, aromatase is encoded by a single gene with a number of tissue-specific promoters. These promoters are under the control of a tissue-specific set of reg-ulatory factors and direct the expression of a number of untranslated 1st exons

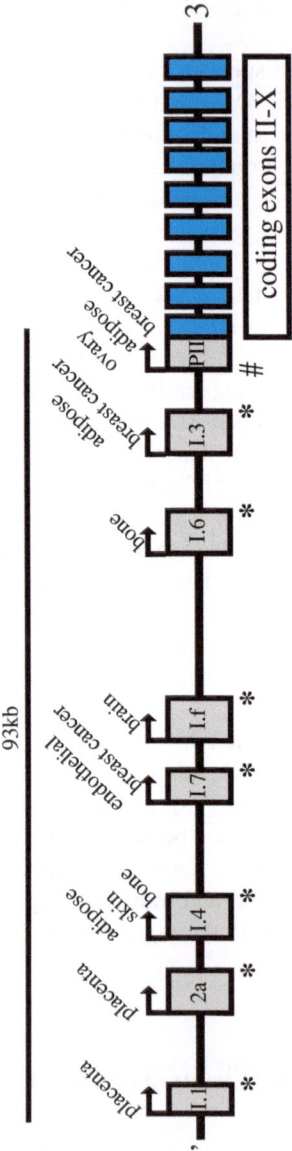

Fig. 4.3 *CYP19A1* gene structure. The human aromatase gene, CYP19A1, is composed of nine coding exons (*blue*) and a number of untranslated first exons (*grey*) which are regulated in a tissue-specific manner. Each first exon has a splice donor site (*) which is spliced into a common splice acceptor site (#) such that coding region is identical irrespective of promoter used to drive expression

which are spliced into the transcript at a common site upstream of the translational start-site such that the protein product is identical in each tissue site of expression (Fig. 4.3). Thus in the placenta, a unique promoter, I.1 is employed which is some 90 kb upstream of the translation start site. This promoter is regulated by factors such as hypoxia factor 1α (HIF1α) and estrogen-related receptor γ (ERRγ) [103]. In adipose tissue, another distal promoter is utilized, I.4, which is regulated by class 1 cytokines and TNFα in the presence of glucocorticoids. However, proximal to a breast tumor, a promoter adjacent to the translation start site is employed, namely promoter PII. This promoter is regulated by cAMP and in the case of adipose tissue, its expression is driven by PGE$_2$. In the ovary, promoter PII is also involved and in this case expression is driven by FSH. The complex nature of the aromatase gene therefore allows the biosynthesis of estrogens at extra-gonadal sites to be finely tuned and responsive to changes in a woman's physiology. Since the initial discovery of human aromatase transcripts in 1986 [104], much work has been done to characterize the structure of the gene that encodes it, *CYP19A1*, and its regulatory regions. In 1988, the full length cDNA from human placental libraries was isolated [105, 106] and the *CYP19A1* gene was mapped to band 15q21.1 of the human genome [107]. In 1989, some of the first regulatory regions of the *CYP19A1* gene were identified by structural analysis, namely a putative TATA sequence located 23 bp upstream of the transcription start site and an AP1 site and cAMP and glucocorticoid regulatory elements [108]. In 1990, the *CYP19A1* gene was found to be greater than 52 kb in size and to consist of 10 exons and 9 introns [109] (Fig. 4.3). The entire gene is now known to encompass some 123 kb, of which, 93 kb is an extended $5'$ regulatory region.

4.5 Promoter-Specific Regulation of Aromatase in Obesity and Breast Cancer

It is plausible that the increase in aromatase expression found in obese breast adipose results in part from the decreased differentiation of stromal cells to adipocytes in response to factors including PGE$_2$ [110], TNFα [111] and IL-11 [112]. This would lead to the increased ratio of adipose stromal cells to fat cells which express little to no aromatase [113, 114]. However, it is clear that factors produced in obesity, including those which inhibit differentiation, alter the expression of aromatase in breast adipose stromal cells (Fig. 4.4) independent of effects on adipogenesis and this, via molecular mechanisms involving complex signaling pathways.

Much of the initial work pertaining to aromatase regulation in the adipose was undertaken before the *CYP19A1* gene structure was even elucidated. In the early 1980s, a number of factors, including glucocorticoids and cAMP analogues, were found to stimulate aromatase activity in isolated human adipose stromal cells [115, 116]. These studies also examined the effect of serum on the glucocorticoid- and cAMP-mediated regulation of aromatase and interestingly, demonstrated that while serum potentiated the effects of dexamethasone, it inhibited the effects of cAMP. This was one of the first indications that glucocorticoids and cAMP must be acting via different pathways to regulate aromatase. Following on from these studies and once the cDNA for aromatase had been isolated, much information was gathered using Northern blot analysis. These studies added a new level of complexity when phorbol esters were found to potentiate the effects of cAMP to increase aromatase expression, and growth factors and inflammatory cytokines, including TNFα and IL-1β, inhibited the cAMP-mediated expression of aromatase [117]. At this stage, interest grew to elucidate the signaling pathways involved. The characterization of the *CYP19A1* gene and the 5'-end of aromatase transcripts suggested that alternative promoters must be used to drive aromatase expression in different tissues, hence the term tissue-specific promoters. It was demonstrated, using 5' rapid amplification of cDNA ends from an adipose-derived cDNA library, that promoter I.3-specific sequences were expressed in adipose tissue as well as in adipose stromal cells maintained under all tissue culture conditions, whereas the newly identified promoter I.4-specific transcripts were present only in breast adipose tissue and adipose stromal cells treated with glucocorticoids [118]. It was later shown that the majority of transcripts found in normal adipose, be it from the buttocks, thighs, abdomen and breast were derived from activation of promoter I.4 [119, 120], whereas adipose tissue from tumor-bearing breast had high levels of aromatase transcripts that were derived from the proximal promoters I.3/II, despite promoter I.4-specific transcripts also being increased [120]. The identification of these specific transcripts allowed additional studies into the promoter-specific regulation of aromatase by tumor-derived and obesity-associated inflammatory factors. In isolated human adipose stromal cells, stimulation with class 1 cytokines, including IL-6, IL-11, leukemia inhibitory factor (LIF) and oncostatin M caused an increase in promoter I.4-specific transcripts, whereas PGE$_2$ stimulated promoter II-specific transcripts [121]. These results are corroborated by findings demonstrating that TNFα, IL-6 and COX-2 are positively correlated with aromatase transcript levels in breast cancer tissue [122].

Both proximal promoters I.3 and II contain a TATA box, yet are also believed to share a number of cis-acting elements. This unique gene organization means that both promoters tend to be coordinately regulated and this accounts for the simultaneous increase in promoter I.3- and II-specific aromatase transcripts observed in cancer-bearing breast tissue and isolated cells in *in vitro* experiments. Conversely, promoter I.4 does not have a TATA or CAAT box upstream of the transcription start site [123], but does require a glucocorticoid response element (GRE) located −133 to −119 to be occupied for activity [123]. The regulation of these promoters, in response to obesity-associated and tumor-derived factors, are detailed below.

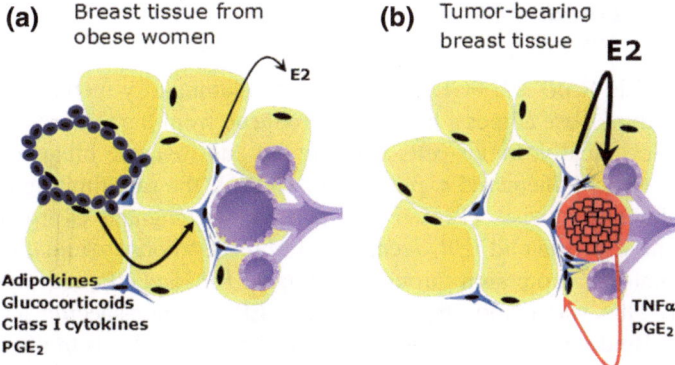

Fig. 4.4 Breast microenvironment in obese women **a** and women with breast cancer **b** Adipose stromal cells (*fibroblast-like blue cells*) have been shown to be the main site of aromatase expression in the human breast. Aromatase expression is increased in obese women as a result of factors produced by adipocytes (*yellow*) and infiltrating immune cells (*round blue cells*). Aromatase expression is also increase as a result of factors produced by the tumor (*red cells*). High aromatase stimulates the proliferation of mammary epithelial cells (*purple*) and breast cancer cells (*red*)

4.5.1 Prostaglandin E_2

As mentioned in Sect. 3.2, PGE_2 is one of the inflammatory mediators which are produced within obese adipose, but is also produced from the tumorous epithelium (Fig. 4.4). Specifically, tumor cells from in situ and invasive breast cancer have been shown to have increased expression of COX-2, the rate-limiting enzyme for prostaglandin biosynthesis [124]. PGE_2 acts via its cognate G-protein-coupled receptors, the E-prostanoid (EP) receptors. In breast adipose stromal cells, PGE_2 induces aromatase expression through binding of EP_1 and EP_2 receptor subtypes [125, 126]. This was demonstrated by using EP receptor-specific agonists and antagonists. Binding of PGE_2 to these prostanoid receptors leads to the subsequent activation of protein kinase A (PKA) and protein kinase C (PKC) (Fig. 4.5). The identification of PGE_2 as a modulator of aromatase expression is not surprising considering early findings demonstrating that cAMP, known to stimulate PKA, and phorbol ester, known to stimulate PKC, act together to increase aromatase expression and activity. The signaling pathways downstream of PKA and PKC have been explored extensively to characterize the PGE_2-mediated induction of aromatase, considered to be a major driver of breast cancer growth. However, emerging research suggests that we have only just scratched the surface in terms of understanding the nature of regulatory complexes and the signaling pathways involved in mediating these effects.

One of the first transcription factors proposed to be involved in the regulation of aromatase promoter I.3/II was the cAMP response element binding protein (CREB). This followed on from results demonstrating that dibutyryl cAMP had stimulatory effects on aromatase expression [116] and the identification of a CRE within the proximal promoter of the aromatase gene [108]. This CRE, located 211–199 bp upstream of the transcription start site, has high sequence homology to a palindromic CRE. However, it also contains an additional cytosine residue and as a consequence, has been referred to as a CRE-like sequence (CLS) [127]. Nevertheless, experiments involving the use of EMSAs revealed that this sequence was capable of binding CREB, albeit with lower affinity than an oligonucleotide containing a CRE, and that mutation of this site significantly reduced the ability of forskolin, which increases intracellular cAMP, to stimulate promoter II in reporter assays [127]. In 1999, a cAMP-response element was also identified and characterized within the promoter I.3 region [128]. This CRE, termed CREaro and containing the sequence TGAAGTCA, is located 66–59 bp upstream of the transcription start site of promoter I.3. EMSAs confirmed the ability of this region to bind to nuclear proteins from both breast tumor-associated stromal cells and the breast cancer cell line SK-BR-3. In 2003, Sofi et al. identified an additional CRE within the promoter II region, located 292–285 bp upstream of the transcription start site which, similar to the CLS site, was shown to also be required for cAMP-induced promoter II activity in mouse 3T3-L1 preadipocytes [129]. EMSAs were also used to confirm binding of CREB to this site using nuclear extracts from 3T3-L1 cells treated with or without forskolin. This same publication also highlighted

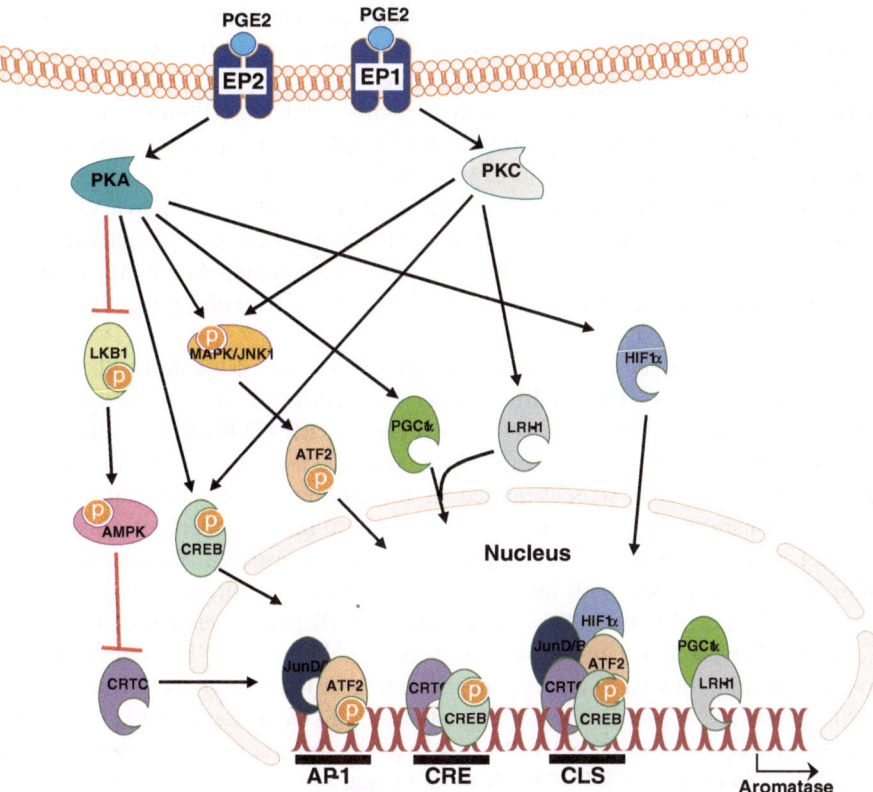

Fig. 4.5 Signaling pathways involved in the regulation of aromatase by PGE$_2$ in breast adipose stromal cells

the biological relevance of CREB in the context of breast cancer by demonstrating that CREB transcript levels were dramatically higher in stromal cells adjacent to tumor tissue than that found in cancer-free breast adipose. CREB is directly phosphorylated by PKA at Ser133 which results in increased affinity for the histone acetylases CREB-binding protein (CBP) and p300. This complex can then bind to CREs and stimulate gene expression. A role for p300 in aromatase regulation was demonstrated using ChIP studies where p300 was shown to interact with aromatase promoter I.3/II and that this interaction was increased in the presence of PGE_2. Using co-IP, it was also demonstrated that PGE_2 treatment caused a noticeable increase in the interaction of p300 with phosphorylated CREB [130].

More recently, other CREB interacting proteins were examined for their role in regulating aromatase. The CREB-regulated transcription coactivators (CRTCs), formally referred to as transducers of regulated CREB (TORCs) and renamed for obvious reasons, are a family of transcriptional coactivators initially identified using high throughput screens aimed at identifying novel modulators of CRE reporter constructs [131]. There are currently three members of this family, CRTC1, CRTC2 and CRTC3, and they all share a conserved N-terminal binding domain which interacts with the bZip domain of CREB. Interestingly, Conkright et al. demonstrated that CRTCs can coactivate CREB independent of whether or not CREB is phosphorylated at Ser133, thereby adding another level of complexity to the regulation of CREB-target genes. The role of the more widely characterized CRTC2 was examined in the context of the PGE_2-dependent regulation of aromatase in adipose stromal cells [132]. It was demonstrated using reporter assays that overexpression of CRTC2 leads to an increase in promoter II activity and this is further increased in the presence of forskolin and phorbol ester, to mimic PGE_2. Moreover, CRTC2 was found to interact with the promoter I.3/II region using chromatin immunoprecipitation on isolated human breast adipose stromal cells, and this interaction is significantly increased in the presence of PGE_2 mimetics, while mutation of the proximal CRE significantly inhibits this effect. The regulation of CRTC2 effects in response to forskolin and phorbol ester was attributed to changes in protein localization. More specifically, CRTC2 was found to be mainly cytoplasmic in resting cells and translocate to the nucleus in the presence of forskolin and phorbol ester. In 2013, Samarajeewa et al. demonstrated that all three CRTCs were capable of increasing aromatase transcript expression, promoter II activity and aromatase enzyme activity and that knockdown of CRTC2 and CRTC3 caused a significant decrease in aromatase activity [133]. The seemingly more important role of CRTC2 and CRTC3 in regulating aromatase may be due to the fact that CRTC1 levels are relatively low in adipose stromal cells. The effects of all three CRTCs also seemed to be additive with CREB and dependent on both CREs for maximal induction of promoter II activity. The established role of CRTC2 in stimulating gluconeogenesis in the liver [134] suggested that metabolic pathways known to regulate CRTC2 activity may also be involved in regulating the CRTC-mediated expression of aromatase in the breast.

One such pathway involves the energy-sensing kinase AMP-activated protein kinase (AMPK), which is considered by many as a master regulator of energy homeostasis. AMPK is active as a heterotrimer, composed of three subunits, namely α, β and γ [reviewed in 135]. In times of low nutrient availability, AMPK is activated and as a consequence inhibits pathways of energy utilisation and stimulates pathways of energy production [136]. AMPK can directly sense energy states as AMP or ADP bind to the regulatory γ subunit of AMPK which leads to conformational changes allowing AMPK to be phosphorylated at the catalytic α subunit at Thr172 [137]. Other phosphorylation sites known to contribute to AMPK regulation include Ser485/491 on the α subunit which leads to the inhibition of AMPK activity [138]. These inhibitory sites have been shown to be phosphorylated by, amongst other kinases, PKA. Two kinases, Liver kinase B1 (LKB1) and calcium/calmodulin-dependent protein kinase kinase β (CaMKKβ), have been shown to act as upstream kinases to AMPK, directly phosphorylating AMPK at Thr172 and leading to its activation. LKB1 is a tumor suppressor and has been shown to be mutated in the majority of cases of the Peutz-Jeghers Syndrome [139]. Affected individuals tend to develop hyperpigmented macules of the oral mucosa and intestinal hamartomatous polyps, and are predisposed to developing a number of cancers including those of the gastrointestinal tract, and other epithelial malignancies including those of the breast [140]. The hypothesis that LKB1/AMPK may be involved in regulating aromatase originated from findings in these patients published in the early 1990s. In addition to the phenotype described above, affected individuals also tend to have symptoms of estrogen excess, including prepubertal gynecomastia and advanced bone age, and two groups independently reported that aromatase expression was increased in the testis of these individuals [141, 142]. Moreover, it was also demonstrated that the majority of aromatase transcripts in these testes were derived from activation of aromatase promoter II [142]. Another clue to this relationship came from findings demonstrating that AMPK negatively regulated CREB-dependent transcription by phosphorylating CRTC2 and preventing its nuclear entry [134]. In order to characterise the role of LKB1/AMPK in aromatase regulation a number of investigations were performed in primary human breast adipose stromal cells [132]. It was demonstrated that LKB1 overexpression was sufficient to inhibit the nuclear translocation of CRTC2 and activation of aromatase promoter II in the presence of forskolin and phorbol ester. It was also demonstrated that the AMPK activator AICAR had a similar inhibitory effect. Interestingly, PGE_2 mimetics were found to decrease LKB1 expression and activity, leading to a decreased phosphorylation of AMPK at Thr172 and the increased nuclear translocation of CRTC2. This was also accompanied by an increase in Ser485/491 phosphorylation. Taken together, this would suggest that under basal conditions, aromatase expression is maintained low in part due to the actions of LKB1/AMPK to inhibit CRTC/CREB activity, whereas in the presence of PGE_2, LKB1/AMPK are suppressed, via mechanisms involving transcript downregulation and increased phosphorylation of inhibitory sites, leading to the translocation of CRTCs, their binding to CREB and activation of aromatase promoter II.

Other kinases involved in the signal transduction downstream of PKA and PKC include the p38 mitogen-activated protein kinase (MAPK) and the c-jun NH2-terminal kinase (JNK) [143, 144]. These kinases have established roles in the regulation of gene expression, cell proliferation and apoptosis [145, 146]. In isolated human breast adipose stromal cells, PGE_2 was found to stimulate the phosphorylation of p38 MAPK and JNK within an hour of treatment [147]. Interestingly, inhibition or silencing of p38 MAPK and JNK1 significantly reduced the PGE_2-mediated expression and activity of aromatase. Moreover, the effects observed were attributed to changes in promoter II and I.3-specific transcripts, whereas inhibition of p38 MAPK and JNK1 had no effect on promoter I.4-specific transcripts. Conversely, overexpression of wild-type p38 MAPK or JNK1 enhanced the PGE_2-mediated induction of aromatase transcript expression. Again, this was attributable to changes in promoter II-specific transcripts and in the case of JNK1 overexpression, also to promoter I.3.

The specific transcriptional complexes hypothesized to be involved in the p38 MAPK/JNK-mediated regulation of aromatase include the Jun family of transcription factors and ATF-2. Jun proteins directly interact with JNK via JNK docking domains thereby leading to their phosphorylation and activation. Jun proteins can form homodimers and heterodimers with other proteins, including ATF-2 [148]. ATF-2 can also be directly phosphorylated and activated by p38 MAPK and JNK. Complexes involving Jun and ATF-2 interact with gene promoters either via binding to activator protein-1 (AP1) or cAMP response elements (CREs). Chromatin immunoprecipitation assays demonstrated that both ATF-2 and c-jun interact with aromatase promoter I.3/II in breast adipose stromal cells in the presence of PGE_2 [147]. However, it was suggested that silencing of ATF-2 had no effect on promoter I.3/II activity whereas silencing of c-jun actually enhanced the expression of promoter I.3/II-specific transcripts [149]. Following on from these studies, Chen et al. demonstrated that other members of the Jun family of transcription factors may in fact be involved in mediating the stimulatory effects of PGE_2 on aromatase expression. In this case, silencing of JunD or JunB in breast adipose stromal cells led to a significant decrease in the PGE_2-mediated expression of total aromatase transcripts, as well as promoter I.3/II-specific transcripts [149]. Interestingly, silencing of JunD stimulated promoter I.4-specific transcript expression. A role for direct interaction of these factors with the proximal CRE (–211/–199) and the AP-1 (–498/–492) binding motifs was demonstrated using DNA precipitation assays, yet only the mutation of the CRE site led to a significant decrease in promoter II activity using reporter assays.

Orphan nuclear receptors SF-1 and LRH-1, encoded by the NR5A1 and NR5A2 genes, respectively, have established roles in the regulation of aromatase in a variety of tissues, including the ovaries and the breast. The first report to demonstrate a role for LRH-1 in breast adipose stromal cells was published by Clyne et al. [113]. Using real-time PCR, it was demonstrated that these cells were deficient in SF-1 and instead expressed LRH-1. Gel shift assays demonstrated that LRH-1 bound to aromatase promoter II at a nuclear receptor half site (AGGTCA) located 130 bp upstream of the transcription start site. Although co-transfection of

LRH-1 with a promoter II reporter construct caused a modest increase in promoter activity in 3T3-L1 cells, used as a model for adipose stromal cells, effects became synergistic in the presence of forskolin and phorbol ester with LRH-1 now increasing promoter II activity by more than 30-fold. A study then followed confirming the clinical relevance of these findings where LRH-1 was shown to be expressed in breast carcinoma tissue as well as the adjacent adipose tissue [150]. A strong correlation between LRH-1 and aromatase mRNA expression was observed in adipose tissue adjacent to breast tumors and this was at least partly attributed to the ability of PGE_2 to stimulate LRH-1 expression in isolated human breast adipose stromal cells. Chromatin immunoprecipitation assays performed using nuclear extracts from adipose stromal cells also confirmed the interaction of LRH-1 with aromatase promoter II and silencing of LRH-1 significantly decreased the ability of forskolin and phorbol ester to increase aromatase expression [151]. The interaction of LRH-1 with promoter II was also shown to involve multiple complexes which modulate its capacity to induce promoter activity. On the one hand, it was shown that the corepressor short heterodimer partner (SHP), expressed under basal conditions in human breast adipose stromal cells, abolished the ability of LRH-1 to increase aromatase promoter II activity in 3T3-L1 cells [152]. Both LRH-1 and SHP are expressed in human stromal cells and cotransfection of 3T3-L1 cells with SHP abolished the basal and forskolin/phorbol ester-stimulated actions of LRH-1 to increase promoter II activity. On the other hand, GATA3/4 [153] and PGC1α [154] have been demonstrated to interact with LRH-1 and synergistically increase aromatase promoter II activity.

Recently, the role of prostaglandins has been expanded to include effects on hypoxia inducible factor 1α (HIF1α) that was until recently, believed to be solely controlled by tissue hypoxia. Indeed, PGE_2 and hypoxia act both independently and synergistically to increase HIF1α in PC-3ML human prostate cancer cells [155] and in HCT116 human colon carcinoma cells [156]. Consistent with these findings, experiments in isolated human breast adipose stromal cells revealed that PGE_2 not only increases HIF1α transcript expression, but also leads to the stabilization of the protein under normoxic conditions [157]. A putative hypoxia response element with sequence 5′-AATGCACGT-3′ was identified in a region of promoter II that overlaps with the proximal CRE. It was found that HIF1α can bind directly to this region. The effect of HIF1α and CREB also appeared to be cooperative and hence implies that the complex which forms in this region of the aromatase promoter may include CREB, CRTC and HIF1α. The absolute requirement for HIF1α for the PGE_2-mediated induction of aromatase was also highlighted by findings demonstrating that silencing of HIF1α was sufficient to abolish the effect of PGE_2. These data were also supported by studies performed on clinical samples. By performing double immunohistochemistry on breast tissue samples from cancer-free and breast cancer patients, a positive association between HIF1α and aromatase was also demonstrated *in vivo*.

The breast cancer susceptibility gene BRCA1 has also been implicated in promoter II/1.3 regulation. In human adipose stromal cells, it was found that knockdown of BRCA1 leads to a significant increase in aromatase transcript expression [158]. A number of additional studies were then undertaken to characterize the role and regulation of BRCA1-dependent inhibition of aromatase in the context of breast cancer. PGE_2 was found to decrease BRCA1 transcript expression in stromal cells, and silencing of BRCA1 in stromal and breast cancer cells led to a significant increase in promoter I.3/II-specific aromatase transcripts [159, 160]. Interestingly, these effects were found to be mediated by EP receptors 2 and 4 [130]. Findings that BRCA1 is a negative regulator of aromatase expression were also confirmed in BRCA1 mutation carriers, where the lack of functional BRCA1 protein correlated with higher promoter II/I.3 and I.4-specific transcripts in breast adipose [161].

In breast epithelial cells, the regulation of aromatase has been found to differ from what is observed in adipose stromal cells. One important difference is the lack of induction of promoter I.3/II in response to activation of CREB, and this may be due to the actions of transcriptional repressors described below. A transcription factor which has been shown to increase epithelial cell aromatase but not stromal cell aromatase, is estrogen-related receptor α (ERRα) [162]. This protein interacts with promoter I.3 in a region called silencer element 1 (S1) located 133–104 bp upstream of the transcription start site and overlaps with the SF-1 element located −136/−124. In SK-BR-3 breast cancer cells, ERRα causes an increase in promoter I.3 activity, however, this transcription factor has no effect on promoter I.3/II activity in 3T3-L1 preadipocytes. In epithelial cells, aromatase expression is suppressed by a number of transcriptional repressors, including COUP-TF1 and EAR-2, which interact with the S1 region [163]. The mechanism of inhibition is likely to involve competition with ERRα for the S1 region. Other transcriptional repressors, which may account for the inability of forskolin to increase aromatase expression in epithelial cells, include SnaH and Slug [164]. These factors bind to a region which overlaps with the distal CRE of aromatase promoter I.3/II and inhibit promoter activity in breast cancer cell lines. Of interest, the expression of these transcriptional repressors was also found to be decreased in breast tumor epithelial cells compared to healthy breast epithelial cells.

4.5.2 Leptin

Leptin has been shown to stimulate aromatase expression in both breast cancer cells and in adipose stromal cells. Catalano et al. demonstrated that in MCF-7 cells, leptin treatment was associated with an increase in aromatase transcript and protein expression as well as activity [165]. The authors demonstrate that leptin acts via MAPK and STAT3 to increase aromatase promoter I.3/II activity, dependent on the AP-1 motif present in the aromatase promoter. In human breast adipose stromal cells, leptin was shown to stimulate aromatase expression with effects being mediated via LKB1/AMPK [132]. More precisely, leptin was shown to decrease LKB1 expression and phosphorylation of AMPK at Thr172, and lead to the increased nuclear localization of CRTC2. This was also associated with an increased binding of CRTC2 to promoter II. In this case, the effects of leptin were similar to those demonstrated for PGE_2 and it is plausible that both factors, increased in obesity, act together to increase aromatase expression and hence estrogen production. Studies using intra-abdominal preadipocytes demonstrated that those from men and women responded differently to leptin treatment [166]. In men, leptin stimulated aromatase transcript expression while no significant effect could be detected in cells isolated from women. Findings in women may be dependent on menopausal status and site of origin. More specifically, cells obtained in the study of Brown et al. were from breast adipose from postmenopausal women whereas cells in the study of Dieudonné et al. were from abdominal adipose depots from premenopausal women.

4.5.3 Adiponectin

Contrary to leptin, adiponectin's effect on aromatase in breast adipose was shown to be inhibitory [132]. This involves the stimulation of AMPK and hence inhibition of CRTC2 nuclear entry and binding to promoter II, thereby leading to the inhibition of aromatase expression. Of interest, adiponectin also stimulated the expression of LKB1 thereby providing a mechanism for the increased activation of AMPK by adiponectin in primary human breast adipose stromal cells. This was also one of the first indications that activation of AMPK was associated with inhibition of aromatase, suggesting that AMPK could be targeted therapeutically to treat hormone receptor positive breast cancer.

4.5.4 IL-6

IL-6 alone and in combination with its soluble receptor (IL-6sR), has been shown to stimulate aromatase activity in isolated adipose stromal cells [167]. In this study, the authors found that IL-6sR caused a 21-fold stimulation in aromatase activity in the presence of IL-6, which was markedly higher than that observed for IL-6 alone. Additional experiments demonstrated that IL-6sR was secreted from MCF7 breast cancer cells, as well as lymphocytes and macrophages, thereby providing evidence for cross-talk between these cell types.

4.5.5 TNFα

TNFα was first shown to stimulate aromatase activity in the presence of dexamethasone in human breast adipose stromal cells in 1994 [168], and later shown to stimulate transcription of aromatase by specifically activating promoter I.4 [169].

In this later study, it was also demonstrated that TNFα caused an increase in c-fos and c-jun protein expression, and that these proteins bound to an imperfect AP-1 binding site found 500–494 bp upstream of the transcription start site within promoter I.4. As mentioned above, binding of the glucocorticoid receptor is absolutely required for activation of promoter I.4 and explains the need to add glucocorticoids such as dexamethasone when stimulating with TNFα. An additional mechanism for the TNFα-mediated regulation of promoter I.4 was recently revealed when it was demonstrated that induction of promoter I.4-specific transcripts by TNFα requires early growth response (Egr) factors in stromal cells [170]. TNFα was shown to induce Egr expression, and this was abrogated in the presence of the MAPK inhibitor U0126, suggesting that NFκB and MAPK are involved in the regulation of Egr proteins by TNFα [171]. The mechanism of aromatase regulation by Egr factors is still unknown. However, these were shown to require a short region of promoter I.4 proximal to the transcription start site, without directly binding to the promoter [170]. A positive feedback loop between estrogens and TNFα signaling has also demonstrated in adipose stromal cells whereby estradiol stimulates the expression of TNF receptors TNFR1 and TNFR2 and in addition to contributing to increased aromatase expression also inhibits adipocyte differentiation [172].

In addition to PGE_2, p38 MAPK and JNK are also activated by TNFα, but unlike PGE2, activation by TNFα does not lead to an increase in aromatase promoter II activity.

4.5.6 Insulin and IGF-1

In 1989, it was demonstrated that insulin had no effect on aromatase activity in human adipose stromal cells in culture either alone or in combination with (Bu)2cAMP [173]. Lueprasitsakul et al. later demonstrated that although insulin and IGF-1 has no effect on aromatase activity on their own, that they markedly attenuated the stimulatory effect of (Bu)2cAMP and significantly increased the dexamethasone-induced activity of aromatase [174]. The much higher concentrations of insulin tested compared to the initial study suggested that insulin and IGF-1 act via the IGF-1 receptor in these cells and account for the discrepancy in results between both studies. Consistent with these findings, a study by Schmidt et al. also demonstrated that insulin increased the cortisol/serum-mediated induction in aromatase activity in breast adipose stromal cells [175]. Of interest, IGF-1 has also been shown to stimulate aromatase activity in MCF-7 and T47D breast cancer cells stably transfected with aromatase, suggesting that IGF-1 may impact posttranslational regulation of aromatase [176].

4.5.7 Other Tumor-Derived Factors

Conditioned medium from the breast cancer cell line T47D has been shown to stimulate aromatase promoter II activity in human adipose stromal cells [177]. This was shown to be cAMP-independent and further studies were undertaken to elucidate the key transcription factors involved. Using deletion constructs of the promoter II reporter, Zhou et al. demonstrated that this induction was dependent on a region located −517 to −278 bp from the transcription start site. Site directed mutagenesis allowed the specific region involved to be pinpointed to a CAAT/enhancer binding protein (C/EBP) binding site located 317–304 bp upstream of the transcription start site. Using EMSA, it was shown that this region interacts with C/EBPβ and C/EBPδ, thereby providing a molecular mechanism for the T47D-mediated induction of promoter II in adipose stromal cells.

Chapter 5
Therapy Aimed at Breaking the Linkage Between Obesity and Breast Cancer

5.1 Effect of Obesity on Endocrine Therapy Efficacy

Current endocrine therapy involving aromatase inhibitors is first line therapy for estrogen receptor positive breast cancer. Higher adiposity is associated with higher aromatase expression hence it has been hypothesized that endocrine therapy may not be as effective in overweight and obese individuals.

When examining the impact of BMI on standard chemohormonal therapy efficacy and patient survival, women who had hormone receptor-positive tumors had a poorer response and a shorter disease free survival compared to HER-2/neu overexpressing or triple negative tumours [178]. Consistent with this, Chen et al. demonstrated that women with a BMI ≥ 25 were 55 % less likely to achieve pathological complete response to neoadjuvant chemotherapy in Chinese women [179]. Interestingly, Sendur et al. demonstrated that BMI had no effect on disease-free survival following aromatase inhibitor use [180]. However, the Breast International Group (BIG) 1–98 study, whereby postmenopausal women received either tamoxifen or letrozole in the adjuvant setting, demonstrated that obese patients had poorer overall survival (HR = 1.19; 95 % CI = 0.99 − 1.44) compared to healthy weight women [181]. There was no difference between the treatment groups.

Nevertheless, Suzuki et al. have recently reviewed the literature relating to the impact of BMI on outcomes of endocrine therapy for women with breast cancer and found that a number of studies suggested that the efficacy of aromatase inhibitors, but not tamoxifen, was affected by BMI [182]. Consistent with this observation, analysis of the ABCSG-06 trial demonstrated that overweight and obese women with breast cancer treated with tamoxifen and the aromatase inhibitor aminoglutethimide had an increased risk of distant recurrences (hazard ratio: 1.67; Cox P = 0.03) and a worse overall survival (hazard ratio: 1.47; Cox P = 0.11) compared with normal weight patients, whereas there was no difference in the different BMI groups for tamoxifen alone [183]. In the ATAC study examining the effect of the non steroidal aromatase inhibitor arimidex versus

K. A. Brown and E. R. Simpson, *Obesity and Breast Cancer*,
SpringerBriefs in Cancer Research, DOI: 10.1007/978-1-4899-8002-1_5,
© The Author(s) 2014

tamoxifen, alone or in combination, women with a high BMI had a significantly worse response to the aromatase inhibitor, whereas efficacy of tamoxifen was not impacted by BMI [184]. Similarly, the BRENDA-cohort study demonstrated obese women tended to benefit more from tamoxifen compared to aromatase inhibitors [185]. Conversely, data obtained from studies using the steroidal aromatase inhibitor exemestane demonstrate that it is more beneficial than tamoxifen in obese patients [182], suggesting that the mechanism by which BMI impacts on aromatase inhibitor efficacy depends on the type of aromatase inhibitor.

When comparing aromatase inhibitors directly, it was demonstrated that the suppression of estradiol and estrone was greater with letrozole than with anastrozole, across all BMI groups [186] and Dorio et al. demonstrated that the impact of aromatase inhibitors on plasma estradiol was not affected by BMI [187].

5.2 Targeting Obesity-Related Estrogen Biosynthesis

5.2.1 Weight Loss, Exercise and Bariatric Surgery: Effect on Breast Cancer Risk and Estrogen Levels

Observational data from recent studies suggests that weight loss reduces the risk for breast cancer [188]. Dietary energy restriction and the prevention of breast cancer is still a topic for debate, however, largely because of difficulties with successfully implementing these types of diets in humans. Nevertheless, convincing data has emerged with short term or intermittent dietary energy restriction in humans and in animal studies. Of note, intermittent dietary restrictions in mice is associated with a reduction in mammary carcinogenesis compared to weight-matched animals which were chronically energy restricted [189], suggesting that the more easily achieved intermittent dietary restriction may be beneficial in humans. Dietary energy restriction has been shown to be associated with beneficial changes in a number of cancer-promoting factors, including insulin sensitivity, glucose availability, IGF-1, sex hormone binding globulin, and estradiol and progesterone [reviewed in 188].

Physical activity on the other hand is associated with improved breast cancer survival. Women meeting physical activity requirements are also less likely to report above-average pain [52] and have a significant reduction in inflammatory markers, including IL-6 [190]. In a review of nine small randomized control trials examining the effect of physical activity on biomarkers in breast cancer survivors, five studies demonstrated a significant beneficial effect on circulating levels of insulin, IGF-I, IGF-II and IGFBPs [191]. It was also noted that inflammatory markers were decreased.

Weight loss has recently been shown to be associated with a decrease in plasma estradiol levels in postmenopausal women (Fig. 5.1). Specifically, in a study by Jones et al. estradiol decreased by 12.7 % for every BMI unit (kg/m^2) lost [192].

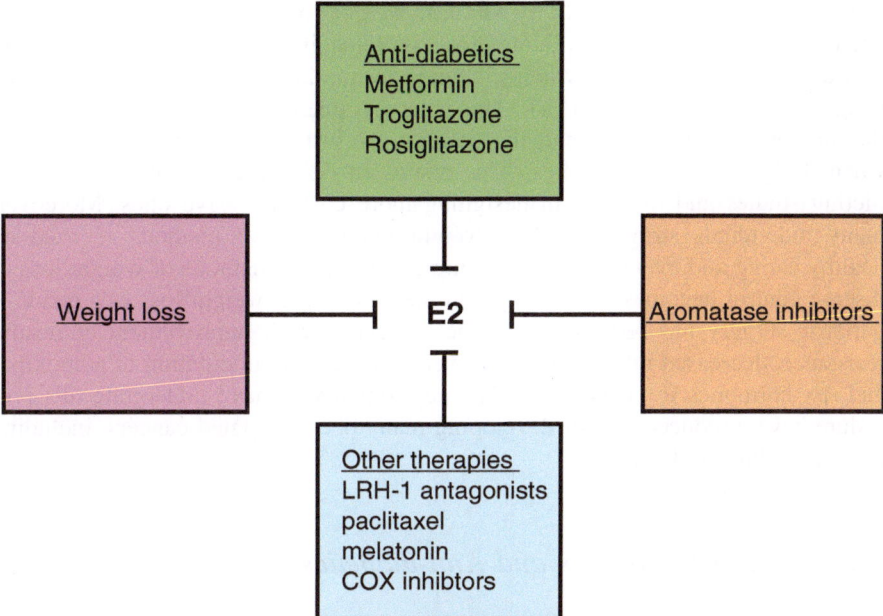

Fig. 5.1 Targeting obesity-related estrogen biosynthesis. Weight loss, anti-diabetics, aromatase inhibitors and a number of other therapies have been shown to decrease estrogen biosynthesis and circulating estrogens, and hence, may prove useful in the treatment of obesity-related postmenopausal breast cancer

Consistent with physical activity studies, weight loss has also been shown to decrease the level of pro-inflammatory proteins in obese women [193–196]. However, there are no convincing studies examining the relationship between weight loss and breast cancer risk. This is largely due to difficulties in ascertaining the amount of weight loss and the duration of sustained weight loss necessary when designing clinical studies. This results in difficulties in interpreting conflicting studies and more so in designing more comprehensive ones. Moreover, many individuals struggle to lose weight via traditional methods focused on healthy eating and exercise, and resorting to alternative methods of weight loss or disease control becomes a necessity. On the other hand, weight loss as a result of bariatric surgery has been shown to be associated with improvement of insulin resistance, decreased inflammation, as well as beneficial modulation of adipokines and sex hormones [reviewed in 197], and women who have undergone this procedure have a reduced risk of developing many obesity-related cancers, including that of the breast [198].

5.2.2 Insulin Sensitizers and Anti-diabetics

The commonly used anti-diabetic drug metformin has received much attention in the past decade for its potential as a cancer therapeutic. This stems largely from observational studies demonstrating that metformin use in diabetics is associated with a significant decrease in the risk of developing a number of cancers, including that of the breast [199, 200]. Use of metformin in the neo-adjuvant setting is associated with a significant decrease in breast tumor proliferation as well as an increase in tumor cell apoptosis. A study by Niraula et al. demonstrated that 500 mg three times/day given to women after their diagnostic biopsy for a medium of 18 days caused a 3 % reduction in Ki67 staining and an almost twofold increase in TUNEL staining [201]. In this study, use of metformin was also associated with a 0.5 kg/m^2 reduction in BMI. Metformin action appears to be mediated by causing changes in ATP levels within cells. This occurs via a number of converging mechanisms including inhibition of complex I in the mitochondrial electron transport chain [reviewed in 202]. The net effect is a lowering of ATP, leading to an increased ratio of AMP to ATP, known to stimulate AMPK. As a consequence of this observation, the use of metformin in settings where AMPK is important, including metabolism, cell proliferation and estrogen biosynthesis, has been explored. Indeed, metformin causes the inhibition of proliferation of a number of endocrine-related cancer cells, including that of the breast [203–206]. In MCF-7 breast cancer cells, treatment with metformin leads to the regulation of a number of genes responsible for cell cycle arrest, including p27Kip1 and p21Cip1 [204]. Metformin has also been shown to inhibit mTOR and as a consequence, decrease translation initiation and protein synthesis [207]. The effect of metformin on cancer cell growth has also been shown to be dependent on the presence of LKB1 as LKB1-deficient cell lines do not respond to metformin treatment.

The effect of metformin on aromatase expression in breast adipose has also recently been described (Fig. 5.1). It was demonstrated in isolated human breast stromal cells, that consistent with its role to activate AMPK in other tissues, metformin stimulated AMPK and caused the cytoplasmic sequestration of CRTC2 in breast stromal cells at micromolar concentrations [208]. This study also provided a novel mechanism for AMPK activation, namely, via the increased expression of LKB1. A subsequent study demonstrated that metformin acted specifically on aromatase promoter I.3/II, with no effect on promoter I.4 [209]. This suggests that metformin may be beneficial in the treatment of hormone receptor positive breast tumors without the side-effects associated with current endocrine therapy use.

Other antidiabetics, including troglitazone and rosiglitazone have also been shown to have an effect on aromatase expression in the breast adipose (Fig. 5.1). In a study by Rubin et al. it was shown that both PPARγ agonists troglitazone and rosiglitazone, which are known to stimulate adipocyte differentiation, inhibit the promoter I.4-driven expression of aromatase [210]. Using Southern blotting, it was demonstrated that these ligands inhibited aromatase transcript expression in oncostatin M or TNFα/dexamethasone-treated primary human breast adipose stromal cells. Moreover, luciferase expression was decreased in aromatase promoter I.4 reporter assays in 3T3-L1 cells treated with troglitazone or rosiglitazone. Of interest, troglitazone also inhibited the forskolin/phorbol ester-mediated expression of aromatase and reporter assays confirmed that effects were dependent on inhibition of aromatase promoter II activity [211]. Nevertheless, EMSAs were unable to demonstrate interactions of PPARγ with the aromatase promoter, suggesting that the effects observed were indirect. This is not unexpected as it is well established that aromatase expression is decreased during the differentiation process.

5.2.3 Other Targeted Therapies

The identification of LRH-1 as a modulator of aromatase, as well as its multiple direct effects on tumor cell growth [212–214] and its tissue-specificity, has led to the hypothesis that inhibition of LRH-1 may lead to breast specific inhibition of estrogen production and cancer cell growth. The search for antagonists or inverse agonists for the receptor has been an ongoing quest, complicated by the fact that the receptor appears to act in the absence of ligand. Recently, Busby et al. identified two inverse agonists for LRH-1, ML179 and ML180, which had IC50 values in the high nanomolar-low micromolar range against aromatase promoter activity [215]. These studies are still in relative infancy but show promise with regards to a new generation of aromatase inhibitors which would act specifically in the breast.

Other factors, including microtubule-stabilizing agents paclitaxel and 2-methoxyestradiol, have been shown to inhibit TNFα, PGE$_2$ and IL-6-mediated aromatase activity in breast stromal cells [216]. Whether these agents also affect bone aromatase is not known. Sodium butyrate has been shown to be promoter selective by decreasing ATF-2 phosphorylation and complex formation with C/EBPdelta and CBP and hence, leads to inhibition of promoters I.3/II but not promoters I.1 or I.4 [217].

Melatonin has also recently been explored for its potential as a breast cancer therapeutic. Higher levels of melatonin have been shown to be inversely associated with breast cancer risk and melatonin levels are known to decrease with aging. The inhibitory effects of melatonin on breast cancer could be explained, at least in part, by the noted inhibitory effects of melatonin on the expression of inflammatory factors TNFα, IL-6 and IL-11, known to stimulate the desmoplastic reaction by inhibiting adipocyte differentiation [218]. A recent study, however, has demonstrated that melatonin can also inhibit the transcription of aromatase driven by promoters I.3/II, as well as promoter I.4, in cancer-associated breast stromal cells [219]. Furthermore, melatonin inhibited the PGE_2-stimulated activity of aromatase in these cells. These results therefore provide an additional mechanism whereby aromatase, and hence estrogen production, is increased in adipose tissue with aging, and suggests that melatonin may be useful as a therapeutic.

Prostaglandins are a major driver of aromatase expression in breast cancer and a number of studies have examined the effect of inhibiting prostaglandin synthesis on aromatase expression and activity. An *in vitro* study by Diaz-Cruz et al. demonstrated that nonsteroidal anti-inflammatory drugs, as well as COX-1 and COX-2 selective inhibitors, potently inhibited aromatase expression and activity in breast cancer cells [220]. A parallel study from the same group demonstrated that COX-1 and COX-2 specific inhibitors suppressed aromatase promoters I.3/II- and I.4-specific aromatase transcripts at micromolar concentrations, but had no effect on promoter I.1-specific transcripts [221].

Conclusions

These studies highlight the possibility of targeting the promoter-specific expression of aromatase and hence the tissue-specific biosynthesis of estrogens. One of our major research focuses remains the identification of therapies which would target breast estrogen formation without affecting estrogen biosynthesis in other tissues, in the bone and brain for example, where estrogens are beneficial. The elucidation of the gene structure and a better understanding of the regulation of aromatase expression in breast tissue has strengthened the rationale for this hypothesis and a number of lead compounds are now being examined clinically. One possibility remains the combination of novel therapies with existing aromatase inhibitors, but in order to obviate the currently experienced side-effects, further studies are required to determine whether a dose-dependent relationship of aromatase inhibitors exists in healthy weight and obese individuals and whether lower doses, alone or in combination, are as effective. Achieving this goal would improve compliance and ameliorate patient quality of life during and after treatment.

K. A. Brown and E. R. Simpson, *Obesity and Breast Cancer*,
SpringerBriefs in Cancer Research, DOI: 10.1007/978-1-4899-8002-1,
© The Author(s) 2014

References

1. Heine, P. A., et al. (2000). Increased adipose tissue in male and female estrogen receptor-alpha knockout mice. *Proceedings of the National Academy of Sciences of the United States of America, 97*(23), 12729–12734.
2. Jones, M. E., et al. (2000). Aromatase-deficient (ArKO) mice have a phenotype of increased adiposity. *Proceedings of the National Academy of Sciences of the United States of America, 97*(23), 12735–12740.
3. Van Sinderen, M. L., et al. (2009). The estrogenic component of tibolone reduces adiposity in female aromatase knockout mice. *Menopause, 16*(3), 582–588.
4. Toth, M. J., et al. (2000). Effect of menopausal status on body composition and abdominal fat distribution. *International Journal of Obesity and Related Metabolic Disorders, 24*(2), 226–231.
5. Teede, H. J., Lombard, C., & Deeks, A. A. (2010). Obesity, metabolic complications and the menopause: An opportunity for prevention. *Climacteric, 13*(3), 203–209.
6. Musatov, S., et al. (2007). Silencing of estrogen receptor alpha in the ventromedial nucleus of hypothalamus leads to metabolic syndrome. *Proceedings of the National Academy of Sciences of the United States of America, 104*(7), 2501–2506.
7. Brown, L. M., & Clegg, D. J. (2010). Central effects of estradiol in the regulation of food intake, body weight, and adiposity. *The Journal of Steroid Biochemistry and Molecular Biology, 122*(1–3), 65–73.
8. Mizutani, T., et al. (1994). Identification of estrogen receptor in human adipose tissue and adipocytes. *The Journal of Clinical Endocrinology and Metabolism, 78*(4), 950–954.
9. Lundholm, L., et al. (2008). Key lipogenic gene expression can be decreased by estrogen in human adipose tissue. *Fertility and Sterility, 90*(1), 44–48.
10. Price, T. M., et al. (1998). Estrogen regulation of adipose tissue lipoprotein lipase–possible mechanism of body fat distribution. *American Journal of Obstetrics and Gynecology, 178*(1 Pt 1), 101–107.
11. McInnes, K. J., et al. (2012). Regulation of LKB1 expression by sex hormones in adipocytes. *International Journal of Obesity (London), 36*(7), 982–985.
12. McInnes, K. J., et al. (2006). Regulation of adenosine 5', monophosphate-activated protein kinase and lipogenesis by androgens contributes to visceral obesity in an estrogen-deficient state. *Endocrinology, 147*(12), 5907–5913.
13. D'Eon, T. M., et al. (2005). Estrogen regulation of adiposity and fuel partitioning. Evidence of genomic and non-genomic regulation of lipogenic and oxidative pathways. *Journal of Biological Chemistry, 280*(43), 35983–35991.
14. Vieira Potter, V. J., et al. (2012). Adipose tissue inflammation and reduced insulin sensitivity in ovariectomized mice occurs in the absence of increased adiposity. *Endocrinology, 153*(9), 4266–4277.

15. Ghisletti, S., et al. (2005). 17beta-estradiol inhibits inflammatory gene expression by controlling NF-kappaB intracellular localization. *Molecular and Cellular Biology, 25*(8), 2957–2968.
16. Appropriate body-mass index for Asian populations and its implications for policy and intervention strategies. (2004). *Lancet 363*(9403), 157–163.
17. World Health Organisation. (2012). *Obesity and overweight*. Fact sheet No 311. Retrieved 2012 from http://www.who.int/mediacentre/factsheets/fs311/en/#.
18. Garrisi, V. M., et al. (2012). Body mass index and serum proteomic profile in breast cancer and healthy women: A prospective study. *PLoS ONE, 7*(11), e49631.
19. Kamineni, A., et al. (2012). Body mass index, tumor characteristics, and prognosis following diagnosis of early-stage breast cancer in a mammographically screened population. *Cancer Causes and Control, 24*, 305–312.
20. Biglia, N., et al. (2012). Body mass index (BMI) and breast cancer: Impact on tumor histopatologic features, cancer subtypes and recurrence rate in pre and postmenopausal women. *Gynecological Endocrinology*.
21. Zhang, F. F., et al. (2012). Total energy intake and breast cancer risk in sisters: The Breast Cancer Family Registry. *Breast Cancer Research and Treatment*
22. van Kruijsdijk, R. C., van der Wall, E., & Visseren, F. L. (2009). Obesity and cancer: The role of dysfunctional adipose tissue. *Cancer Epidemiology, Biomarkers and Prevention, 18*(10), 2569–2578.
23. Cheraghi, Z., et al. (2012). Effect of body mass index on breast cancer during premenopausal and postmenopausal periods: A meta-analysis. *PLoS ONE, 7*(12), e51446.
24. Bergstrom, A., et al. (2001). Overweight as an avoidable cause of cancer in Europe. *International Journal of Cancer, 91*(3), 421–430.
25. Ursin, G., et al. (1995). A meta-analysis of body mass index and risk of premenopausal breast cancer. *Epidemiology, 6*(2), 137–141.
26. Lahmann, P. H., et al. (2004). Body size and breast cancer risk: Findings from the European Prospective Investigation into Cancer and Nutrition (EPIC). *International Journal of Cancer, 111*(5), 762–771.
27. Weiderpass, E., et al. (2004). A prospective study of body size in different periods of life and risk of premenopausal breast cancer. *Cancer Epidemiology, Biomarkers and Prevention, 13*(7), 1121–1127.
28. Michels, K. B., Terry, K. L., & Willett, W. C. (2006). Longitudinal study on the role of body size in premenopausal breast cancer. *Archives of Internal Medicine, 166*(21), 2395–2402.
29. Pierobon, M., & Frankenfeld, C. L. (2013). Obesity as a risk factor for triple-negative breast cancers: A systematic review and meta-analysis. *Breast Cancer Research and Treatment, 137*(1), 307–314.
30. Ricks-Santi, L. J., et al. (2013). BRCA1 polymorphisms and breast cancer epidemiology in the Western New York Exposures and Breast cancer (WEB) study. *Genetic Epidemiology, 37*(5), 504–511.
31. Key, T. J., et al. (2003). Body mass index, serum sex hormones, and breast cancer risk in postmenopausal women. *The Journal of the National Cancer Institute, 95*(16), 1218–26.
32. Kimura, K., et al. (2012). Association between body mass index and breast cancer intrinsic subtypes in Japanese women. *Experimental and Therapeutic Medicine, 4*(3), 391–396.
33. John, E. M., Phipps, A. I., & Sangaramoorthy, M. (2013). Body size, modifying factors, and postmenopausal breast cancer risk in a multiethnic population: The San Francisco bay area breast cancer study. *Springerplus, 2*(1), 239.
34. Krishnan, K., et al. (2013). Associations between weight in early adulthood, change in weight and breast cancer risk in postmenopausal women. *Cancer Epidemiology Biomarkers & Prevention*.
35. Boyd, N. F., et al. (2011). Mammographic density and breast cancer risk: Current understanding and future prospects. *Breast Cancer Research, 13*(6), 223.

36. Dorgan, J. F., et al. (2012). Height, adiposity and body fat distribution and breast density in young women. *Breast Cancer Research, 14*(4), R107.
37. Boyd, N. F., et al. (1998). The relationship of anthropometric measures to radiological features of the breast in premenopausal women. *British Journal of Cancer, 78*(9), 1233–8.
38. Vachon, C. M., et al. (2010). Aromatase immunoreactivity is increased in mammographically dense regions of the breast. *Breast Cancer Research and Treatment, 125*, 243–252 (Epub ahead of print).
39. Fuhrman, B. J., et al. (2012). Estrogen metabolism and mammographic density in postmenopausal women: A cross-sectional study. *Cancer Epidemiology, Biomarkers and Prevention, 21*(9), 1582–1591.
40. Amadou, A., et al. (2013). Overweight, obesity and risk of premenopausal breast cancer according to ethnicity: A systematic review and dose-response meta-analysis. *Obesity Reviews*.
41. Abdullah, A., et al. (2010). The magnitude of association between overweight and obesity and the risk of diabetes: A meta-analysis of prospective cohort studies. *Diabetes Research and Clinical Practice, 89*(3), 309–319.
42. Forte, V., et al. (2012). Obesity, diabetes, the cardiorenal syndrome, and risk for cancer. *Cardiorenal Medicine, 2*(2), 143–162.
43. Almdal, T., et al. (2004). The independent effect of type 2 diabetes mellitus on ischemic heart disease, stroke, and death: A population-based study of 13,000 men and women with 20 years of follow-up. *Archives of Internal Medicine, 164*(13), 1422–1426.
44. Seshasai, S. R., et al. (2011). Diabetes mellitus, fasting glucose, and risk of cause-specific death. *New England Journal of Medicine, 364*(9), 829–841.
45. Boyle, P., et al. (2012). Diabetes and breast cancer risk: A meta-analysis. *British Journal of Cancer, 107*(9), 1608–1617.
46. Yu, Z. G., et al. (2012). The prevalence and correlates of breast cancer among women in Eastern China. *PLoS ONE, 7*(6), e37784.
47. Esposito, K., et al. (2013). Metabolic syndrome and postmenopausal breast cancer: Systematic review and meta-analysis. *Menopause*.
48. Kusano, A. S., et al. (2006). A prospective study of breast size and premenopausal breast cancer incidence. *International Journal of Cancer, 118*(8), 2031–2034.
49. Markkula, A., et al. (2012). Given breast cancer, does breast size matter? Data from a prospective breast cancer cohort. *Cancer Causes and Control, 23*(8), 1307–1316.
50. Druesne-Pecollo, N., et al. (2012). Excess body weight and second primary cancer risk after breast cancer: A systematic review and meta-analysis of prospective studies. *Breast Cancer Research and Treatment, 135*(3), 647–654.
51. Liu, Y. L., et al. (2012). Association of serum lipid profile with distant metastasis in breast cancer patients. *Zhonghua Zhong Liu Za Zhi, 34*(2), 129–131.
52. Forsythe, L. P., et al. (2012). Pain in long-term breast cancer survivors: The role of body mass index, physical activity, and sedentary behavior. *Breast Cancer Research and Treatment*.
53. Campbell, P. T., et al. (2012). Diabetes and cause-specific mortality in a prospective cohort of one million U.S. adults. *Diabetes Care, 35*(9), 1835–1844.
54. Lehr, S., et al. (2012). Identification and validation of novel adipokines released from primary human adipocytes. *Molecular & Cellular Proteomics, 11*(1), M111 010504.
55. Ryan, A. S., et al. (2003). Plasma adiponectin and leptin levels, body composition, and glucose utilization in adult women with wide ranges of age and obesity. *Diabetes Care, 26*(8), 2383–2388.
56. Kelesidis, I., Kelesidis, T., & Mantzoros, C. S. (2006). Adiponectin and cancer: A systematic review. *British Journal of Cancer, 94*(9), 1221–1225.
57. Gulcelik, M. A., et al. (2012). Associations between adiponectin and two different cancers: Breast and colon. *Asian Pacific Journal of Cancer Prevention, 13*(1), 395–398.

58. Rajala, M. W., & Scherer, P. E. (2003). Minireview: The adipocyte—at the crossroads of energy homeostasis, inflammation, and atherosclerosis. *Endocrinology, 144*(9), 3765–3773.
59. Rose, D. P., Komninou, D., & Stephenson, G. D. (2004). Obesity, adipocytokines, and insulin resistance in breast cancer. *Obesity Review, 5*(3), 153–165.
60. Wu, M.-H., et al. (2009). Circulating levels of leptin, adiposity and breast cancer risk. *British Journal of Cancer, 100*, 578–582.
61. Maccio, A., et al. (2010). Correlation of body mass index and leptin with tumor size and stage of disease in hormone-dependent postmenopausal breast cancer: Preliminary results and therapeutic implications. *Journal of Molecular Medicine (Berlin), 88*(7), 677–686.
62. Grossmann, M. E., et al. (2010). Obesity and breast cancer: Status of leptin and adiponectin in pathological processes. *Cancer and Metastasis Reviews, 29*(4), 641–653.
63. Cirillo, D., et al. (2008). Leptin signaling in breast cancer: An overview. *Journal of Cellular Biochemistry, 105*(4), 956–964.
64. Jarde, T., et al. (2011). Molecular mechanisms of leptin and adiponectin in breast cancer. *European Journal of Cancer, 47*(1), 33–43.
65. Xue, R. Q., et al. (2012). Effects of exogenous human leptin on heat shock protein 70 expression in MCF-7 breast cancer cells and breast carcinoma of nude mice xenograft model. *Chinese Medical Journal (English Edition), 125*(4), 680–686.
66. Barnes, J. A., et al. (2001). Expression of inducible Hsp70 enhances the proliferation of MCF-7 breast cancer cells and protects against the cytotoxic effects of hyperthermia. *Cell Stress and Chaperones, 6*(4), 316–325.
67. Zheng, Q., Hursting, S. D., & Reizes, O. (2012). Leptin regulates cyclin D1 in luminal epithelial cells of mouse MMTV-Wnt-1 mammary tumors. *Journal of Cancer Research and Clinical Oncology, 138*(9), 1607–1612.
68. Dalamaga, M. (2012). Nicotinamide phosphoribosyl-transferase/visfatin: A missing link between overweight/obesity and postmenopausal breast cancer? Potential preventive and therapeutic perspectives and challenges. *Medical Hypotheses, 79*(5), 617–621.
69. Lee, Y. C., et al. (2012). Resistin expression in breast cancer tissue as a marker of prognosis and hormone therapy stratification. *Gynecologic Oncology, 125*(3), 742–750.
70. Llanos, A. A., et al. (2012). Adipokines in plasma and breast tissues: Associations with breast cancer risk factors. *Cancer Epidemiology, Biomarkers & Prevention, 21*(10), 1745–1755.
71. Simons, P. J., et al. (2005). Cytokine-mediated modulation of leptin and adiponectin secretion during in vitro adipogenesis: Evidence that tumor necrosis factor-alpha- and interleukin-1beta-treated human preadipocytes are potent leptin producers. *Cytokine, 32*(2), 94–103.
72. Vona-Davis, L., & Rose, D. P. (2009). Angiogenesis, adipokines and breast cancer. *Cytokine & Growth Factor Reviews, 20*(3), 193–201.
73. Zeyda, M., et al. (2010). Newly identified adipose tissue macrophage populations in obesity with distinct chemokine and chemokine receptor expression. *International Journal of Obesity (London), 34*(12), 1684–1694.
74. Subbaramaiah, K., et al. (2011). Obesity is associated with inflammation and elevated aromatase expression in the mouse mammary gland. *Cancer Prevention Research (Philadelphia), 4*(3), 329–346.
75. Subbaramaiah, K., et al. (2012). Increased levels of COX-2 and prostaglandin E2 contribute to elevated aromatase expression in inflamed breast tissue of obese women. *Cancer Discovery, 2*(4), 356–365.
76. Chen, J. (2011). Multiple signal pathways in obesity-associated cancer. *Obesity Reviews, 12*(12), 1063–1070.
77. Cannata, D., et al. (2010). Type 2 diabetes and cancer: What is the connection? *Mount Sinai Journal of Medicine, 77*(2), 197–213.

78. Chong, Y. M., et al. (2006). The relationship between the insulin-like growth factor-1 system and the oestrogen metabolising enzymes in breast cancer tissue and its adjacent non-cancerous tissue. *Breast Cancer Research and Treatment, 99*(3), 275–288.

79. Rice, M. S., et al. (2012). Insulin-like growth factor-1, insulin-like growth factor-binding protein-3, growth hormone, and mammographic density in the Nurses' Health Studies. *Breast Cancer Research and Treatment, 136*(3), 805–812.

80. McNatty, K. P., et al. (1976). Concentration of oestrogens and androgens in human ovarian venous plasma and follicular fluid throughout the menstrual cycle. *Journal of Endocrinology, 71*(1), 77–85.

81. Fournet-Dulguerov, N., et al. (1987). Immunohistochemical localization of aromatase cytochrome P-450 and estradiol dehydrogenase in the syncytiotrophoblast of the human placenta. *Journal of Clinical Endocrinology and Metabolism, 65*(4), 757–764.

82. Grodin, J. M., Siiteri, P. K., & MacDonald, P. C. (1973). Source of estrogen production in postmenopausal women. *Journal of Clinical Endocrinology and Metabolism, 36*(2), 207–214.

83. Hemsell, D. L., et al. (1974). Plasma precursors of estrogen. II. Correlation of the extent of conversion of plasma androstenedione to estrone with age. *Journal of Clinical Endocrinology and Metabolism, 38*(3), 476–479.

84. Edman, C. D., & MacDonald, P. C. (1978). Effect of obesity on conversion of plasma androstenedione to estrone in ovulatory and anovulator young women. *American Journal of Obstetrics and Gynecology, 130*(4), 456–461.

85. Miller, W. R., & O'Neill, J. (1987). The importance of local synthesis of estrogen within the breast. *Steroids, 50*(4–6), 537–548.

86. Nimrod, A., & Ryan, K. J. (1975). Aromatization of androgens by human abdominal and breast fat tissue. *Journal of Clinical Endocrinology and Metabolism, 40*(3), 367–372.

87. Thompson, E. A., Jr. & Siiteri, P. K., (1974) Utilization of oxygen and reduced nicotinamide adenine dinucleotide phosphate by human placental microsomes during aromatization of androstenedione. *Journal of Biological Chemistry, 249*(17), 5364–5372.

88. Ghosh, D., et al. (2009). Structural basis for androgen specificity and oestrogen synthesis in human aromatase. *Nature, 457*(7226), 219–223.

89. Hankinson, S. E., et al. (1998). Plasma sex steroid hormone levels and risk of breast cancer in postmenopausal women. *Journal of the National Cancer Institute, 90*(17), 1292–1299.

90. Huang, Z., et al. (1997). Dual effects of weight and weight gain on breast cancer risk. *The Journal of the American Medical Association, 278*(17), 1407–1411.

91. Haynes, B. P., et al. (2010). Intratumoral estrogen disposition in breast cancer. *Clinical Cancer Research, 16*(6), 1790–1801.

92. O'Neill, J. S., Elton, R. A., & Miller, W. R. (1988). Aromatase activity in adipose tissue from breast quadrants: A link with tumour site. *British Medical Journal (Clinical Research Ed), 296*(6624), 741–743.

93. Bulun, S. E., et al. (1993). A link between breast cancer and local estrogen biosynthesis suggested by quantification of breast adipose tissue aromatase cytochrome P450 transcripts using competitive polymerase chain reaction after reverse transcription. *Journal of Clinical Endocrinology and Metabolism, 77*(6), 1622–1628.

94. Utsumi, T., et al. (1996). Presence of alternatively spliced transcripts of aromatase gene in human breast cancer. *Journal of Clinical Endocrinology and Metabolism, 81*(6), 2344–2349.

95. Zhou, C., et al. (1996). Aromatase gene expression and its exon I usage in human breast tumors. Detection of aromatase messenger RNA by reverse transcription-polymerase chain reaction. *Journal of Steroid Biochemistry and Molecular Biology, 59*(2), 163–171.

96. Sasano, H., et al. (1994). Immunolocalization of aromatase and other steroidogenic enzymes in human breast disorders. *Human Pathology, 25*(5), 530–535.

97. Harada, N., Utsumi, T., & Takagi, Y. (1993). Tissue-specific expression of the human aromatase cytochrome P-450 gene by alternative use of multiple exons 1 and promoters, and switching of tissue-specific exons 1 in carcinogenesis. *Proceedings of the National Academy of Sciences of the United States of America, 90*(23), 11312–11316.

98. Tekmal, R. R., et al. (1999). Aromatase overexpression and breast hyperplasia, an in vivo model–continued overexpression of aromatase is sufficient to maintain hyperplasia without circulating estrogens, and aromatase inhibitors abrogate these preneoplastic changes in mammary glands. *Endocrine-Related Cancer, 6*(2), 307–314.

99. Ackerman, G. E., et al. (1981). Aromatization of androstenedione by human adipose tissue stromal cells in monolayer culture. *Journal of Clinical Endocrinology and Metabolism, 53*(2), 412–417.

100. Santen, R. J., et al. (1994). Stromal spindle cells contain aromatase in human breast tumors. *Journal of Clinical Endocrinology and Metabolism, 79*(2), 627–632.

101. Sasano, H., et al. (1996). Aromatase and 17 beta-hydroxysteroid dehydrogenase type 1 in human breast carcinoma. *Journal of Clinical Endocrinology and Metabolism, 81*(11), 4042–4046.

102. Brodie, A., Long, B., & Lu, Q., (1998). Aromatase expression in the human breast. *Breast Cancer Research and Treatment, 49*(Suppl 1), S85–S91 (discussion S109–19).

103. Kumar, P., & Mendelson, C. R. (2011). Estrogen-related receptor gamma (ERRgamma) mediates oxygen-dependent induction of aromatase (CYP19) gene expression during human trophoblast differentiation. *Molecular Endocrinology, 25*(9), 1513–1526.

104. Evans, C. T., et al. (1986). Isolation and characterization of a complementary DNA specific for human aromatase-system cytochrome P-450 mRNA. *Proceedings of the National Academy of Sciences of the United States of America, 83*(17), 6387–6391.

105. Corbin, C. J., et al. (1988). Isolation of a full-length cDNA insert encoding human aromatase system cytochrome P-450 and its expression in nonsteroidogenic cells. *Proceedings of the National Academy of Sciences of the United States of America, 85*(23), 8948–8952.

106. Harada, N. (1988). Cloning of a complete cDNA encoding human aromatase: Immunochemical identification and sequence analysis. *Biochemical and Biophysical Research Communications, 156*(2), 725–732.

107. Chen, S. A., et al. (1988). Human aromatase: cDNA cloning, Southern blot analysis, and assignment of the gene to chromosome 15. *DNA, 7*(1), 27–38.

108. Means, G. D., et al. (1989). Structural analysis of the gene encoding human aromatase cytochrome P-450, the enzyme responsible for estrogen biosynthesis. *Journal of Biological Chemistry, 264*(32), 19385–19391.

109. Mendelson, C. R., et al. (1990). Use of molecular probes to study regulation of aromatase cytochrome P-450. *Biology of Reproduction, 42*(1), 1–10.

110. Tsuboi, H., et al. (2004). Prostanoid EP4 receptor is involved in suppression of 3T3-L1 adipocyte differentiation. *Biochemical and Biophysical Research Communications, 322*(3), 1066–1072.

111. Kawakami, M., et al. (1989). Cachectin/TNF kills or inhibits the differentiation of 3T3-L1 cells according to developmental stage. *Journal of Cellular Physiology, 138*(1), 1–7.

112. Kawashima, I., et al. (1991). Molecular cloning of cDNA encoding adipogenesis inhibitory factor and identity with interleukin-11. *FEBS Letters, 283*(2), 199–202.

113. Clyne, C. D., et al. (2002). Liver receptor homologue-1 (LRH-1) regulates expression of aromatase in preadipocytes. *Journal of Biological Chemistry, 277*(23), 20591–20597.

114. Dieudonne, M. N., et al. (2006). Sex steroids and leptin regulate 11beta-hydroxysteroid dehydrogenase I and P450 aromatase expressions in human preadipocytes: Sex specificities. *Journal of Steroid Biochemistry and Molecular Biology, 99*(4–5), 189–196.

115. Simpson, E. R., et al. (1981). Estrogen formation in stromal cells of adipose tissue of women: Induction by glucocorticosteroids. *Proceedings of the National Academy of Sciences of the United States of America, 78*(9), 5690–5694.

116. Mendelson, C. R., et al. (1982). Regulation of aromatase activity of stromal cells derived from human adipose tissue. *Endocrinology, 111*(4), 1077–1085.

117. Simpson, E. R., et al. (1989). Regulation of estrogen biosynthesis by human adipose cells. *Endocrine Reviews, 10*(2), 136–148.

118. Mahendroo, M. S., Mendelson, C. R., & Simpson, E. R. (1993). Tissue-specific and hormonally controlled alternative promoters regulate aromatase cytochrome P450 gene expression in human adipose tissue. *Journal of Biological Chemistry, 268*(26), 19463–19470.

119. Agarwal, V. R., et al. (1997). Alternatively spliced transcripts of the aromatase cytochrome P450 (CYP19) gene in adipose tissue of women. *Journal of Clinical Endocrinology and Metabolism, 82*(1), 70–74.

120. Agarwal, V. R., et al. (1996). Use of alternative promoters to express the aromatase cytochrome P450 (CYP19) gene in breast adipose tissues of cancer-free and breast cancer patients. *Journal of Clinical Endocrinology and Metabolism, 81*(11), 3843–3849.

121. Zhao, Y., et al. (1997). Transcriptional regulation of CYP19 gene (aromatase) expression in adipose stromal cells in primary culture. *Journal of Steroid Biochemistry and Molecular Biology, 61*(3–6), 203–210.

122. Irahara, N., et al. (2006). Quantitative analysis of aromatase mRNA expression derived from various promoters (I.4, I.3, PII and I.7) and its association with expression of TNF-alpha, IL-6 and COX-2 mRNAs in human breast cancer. *International Journal of Cancer, 118*(8), 1915–1921.

123. Zhao, Y., Mendelson, C. R., & Simpson, E. R. (1995). Characterization of the sequences of the human CYP19 (aromatase) gene that mediate regulation by glucocorticoids in adipose stromal cells and fetal hepatocytes. *Molecular Endocrinology, 9*(3), 340–349.

124. Half, E., et al. (2002). Cyclooxygenase-2 expression in human breast cancers and adjacent ductal carcinoma in situ. *Cancer Research, 62*(6), 1676–1681.

125. Richards, J. A., & Brueggemeier, R. W. (2003). Prostaglandin E2 regulates aromatase activity and expression in human adipose stromal cells via two distinct receptor subtypes. *Journal of Clinical Endocrinology and Metabolism, 88*(6), 2810–2816.

126. Zhao, Y., et al. (1996). Estrogen biosynthesis proximal to a breast tumor is stimulated by PGE2 via cyclic AMP, leading to activation of promoter II of the CYP19 (aromatase) gene. *Endocrinology, 137*(12), 5739–5742.

127. Michael, M. D., Michael, L. F., & Simpson, E. R. (1997). A CRE-like sequence that binds CREB and contributes to cAMP-dependent regulation of the proximal promoter of the human aromatase P450 (CYP19) gene. *Molecular and Cellular Endocrinology, 134*(2), 147–156.

128. Zhou, D., & Chen, S. (1999). Identification and characterization of a cAMP-responsive element in the region upstream from promoter 1.3 of the human aromatase gene. *Archives of Biochemistry and Biophysics, 371*(2), 179–190.

129. Sofi, M., et al. (2003). Role of CRE-binding protein (CREB) in aromatase expression in breast adipose. *Breast Cancer Research and Treatment, 79*(3), 399–407.

130. Subbaramaiah, K., et al. (2008). EP2 and EP4 receptors regulate aromatase expression in human adipocytes and breast cancer cells. Evidence of a BRCA1 and p300 exchange. *Journal of Biological Chemistry, 283*(6), 3433–3444.

131. Conkright, M. D., et al. (2003). TORCs: Transducers of regulated CREB activity. *Molecular Cell, 12*(2), 413–23.

132. Brown, K. A., et al. (2009). Subcellular localization of cyclic AMP-responsive element binding protein-regulated transcription coactivator 2 provides a link between obesity and breast cancer in postmenopausal women. *Cancer Research, 69*(13), 5392–5399.

133. Samarajeewa, N. U., et al. (2013) CREB-regulated transcription co-activator family stimulates promoter II-driven aromatase expression in preadipocytes. *Hormones and Cancer*.

134. Koo, S. H., et al. (2005). The CREB coactivator TORC2 is a key regulator of fasting glucose metabolism. *Nature, 437*(7062), 1109–1111.
135. Steinberg, G. R., & Kemp, B. E. (2009). AMPK in health and disease. *Physiological Reviews, 89*(3), 1025–1078.
136. Towler, M. C., & Hardie, D. G. (2007). AMP-activated protein kinase in metabolic control and insulin signaling. *Circulation Research, 100*(3), 328–341.
137. Oakhill, J. S., et al. (2011). AMPK is a direct adenylate charge-regulated protein kinase. *Science, 332*(6036), 1433–1435.
138. Hurley, R. L., et al. (2006). Regulation of AMP-activated protein kinase by multisite phosphorylation in response to agents that elevate cellular cAMP. *Journal of Biological Chemistry, 281*(48), 36662–36672.
139. Hemminki, A., et al. (1998). A serine/threonine kinase gene defective in Peutz-Jeghers syndrome. *Nature, 391*(6663), 184–187.
140. Alessi, D. R., Sakamoto, K., & Bayascas, J. R. (2006). LKB1-dependent signaling pathways. *Annual Review of Biochemistry, 75*, 137–163.
141. Coen, P., et al. (1991). An aromatase-producing sex-cord tumor resulting in prepubertal gynecomastia. *New England Journal of Medicine, 324*(5), 317–322.
142. Bulun, S. E., et al. (1994). Use of tissue-specific promoters in the regulation of aromatase cytochrome P450 gene expression in human testicular and ovarian sex cord tumors, as well as in normal fetal and adult gonads. *Journal of Clinical Endocrinology and Metabolism, 78*(2), 1616–1621.
143. Schonwasser, D. C., et al. (1998). Activation of the mitogen-activated protein kinase/extracellular signal-regulated kinase pathway by conventional, novel, and atypical protein kinase C isotypes. *Molecular and Cellular Biology, 18*(2), 790–798.
144. Zhang, B., et al. (2004). JNK signaling involved in the effects of cyclic AMP on IL-1beta plus IFNgamma-induced inducible nitric oxide synthase expression in hepatocytes. *Cellular Signalling, 16*(7), 837–846.
145. Davis, R. J. (2000). Signal transduction by the JNK group of MAP kinases. *Cell, 103*(2), 239–252.
146. Roux, P. P., & Blenis, J. (2004). ERK and p38 MAPK-activated protein kinases: A family of protein kinases with diverse biological functions. *Microbiology and Molecular Biology Reviews, 68*(2), 320–344.
147. Chen, D., et al. (2007). Prostaglandin E(2) induces breast cancer related aromatase promoters via activation of p38 and c-Jun NH(2)-terminal kinase in adipose fibroblasts. *Cancer Research, 67*(18), 8914–8922.
148. Mechta-Grigoriou, F., Gerald, D., & Yaniv, M. (2001). The mammalian Jun proteins: Redundancy and specificity. *Oncogene, 20*(19), 2378–2389.
149. Chen, D., et al. (2011). JunD and JunB integrate prostaglandin E2 activation of breast cancer-associated proximal aromatase promoters. *Molecular Endocrinology, 25*(5), 767–775.
150. Zhou, J., et al. (2005). Interactions between prostaglandin E(2), liver receptor homologue-1, and aromatase in breast cancer. *Cancer Research, 65*(2), 657–663.
151. Chand, A. L., et al. (2011). Tissue-specific regulation of aromatase promoter II by the orphan nuclear receptor LRH-1 in breast adipose stromal fibroblasts. *Steroids, 76*(8), 741–744.
152. Kovacic, A., et al. (2004). Inhibition of aromatase transcription via promoter II by short heterodimer partner in human preadipocytes. *Molecular Endocrinology, 18*(1), 252–259.
153. Bouchard, M. F., Taniguchi, H., & Viger, R. S. (2005). Protein kinase A-dependent synergism between GATA factors and the nuclear receptor, liver receptor homolog-1, regulates human aromatase (CYP19) PII promoter activity in breast cancer cells. *Endocrinology, 146*(11), 4905–4916.

154. Safi, R., et al. (2005). Coactivation of liver receptor homologue-1 by peroxisome proliferator-activated receptor gamma coactivator-1alpha on aromatase promoter II and its inhibition by activated retinoid X receptor suggest a novel target for breast-specific antiestrogen therapy. *Cancer Research, 65*(24), 11762–11770.

155. Liu, X. H., et al. (2002). Prostaglandin E2 induces hypoxia-inducible factor-1alpha stabilization and nuclear localization in a human prostate cancer cell line. *Journal of Biological Chemistry, 277*(51), 50081–50086.

156. Fukuda, R., Kelly, B., & Semenza, G. L. (2003). Vascular endothelial growth factor gene expression in colon cancer cells exposed to prostaglandin E2 is mediated by hypoxia-inducible factor 1. *Cancer Research, 63*(9), 2330–2334.

157. Samarajeewa, N. U., et al. (2013). HIF-1alpha stimulates aromatase expression driven by prostaglandin E2 in breast adipose stroma. *Breast Cancer Research, 15*(2), R30.

158. Hu, Y., et al. (2005). Modulation of aromatase expression by BRCA1: A possible link to tissue-specific tumor suppression. *Oncogene, 24*(56), 8343–8348.

159. Lu, M., et al. (2006). BRCA1 negatively regulates the cancer-associated aromatase promoters I.3 and II in breast adipose fibroblasts and malignant epithelial cells. *Journal of Clinical Endocrinology and Metabolism, 91*(11), 4514–4519.

160. Ghosh, S., et al. (2007). Tumor suppressor BRCA1 inhibits a breast cancer-associated promoter of the aromatase gene (CYP19) in human adipose stromal cells. *American Journal of Physiology: Endocrinology and Metabolism, 292*(1), E246–E252.

161. Chand, A. L., Simpson, E. R., & Clyne, C. D. (2009). Aromatase expression is increased in BRCA1 mutation carriers. *BMC Cancer, 9*, 148.

162. Yang, C., Zhou, D., & Chen, S. (1998). Modulation of aromatase expression in the breast tissue by ERR alpha-1 orphan receptor. *Cancer Research, 58*(24), 5695–5700.

163. Yang, C., et al. (2002). Regulation of aromatase promoter activity in human breast tissue by nuclear receptors. *Oncogene, 21*(18), 2854–2863.

164. Okubo, T., et al. (2001). Down-regulation of promoter 1.3 activity of the human aromatase gene in breast tissue by zinc-finger protein, snail (SnaH). *Cancer Research, 61*(4), 1338–1346.

165. Catalano, S., et al. (2003). Leptin enhances, via AP-1, expression of aromatase in the MCF-7 cell line. *Journal of Biological Chemistry, 278*(31), 28668–28676.

166. Dieudonne, M. N., et al. (2006). Adiponectin mediates antiproliferative and apoptotic responses in human MCF7 breast cancer cells. *Biochemical and Biophysical Research Communications, 345*(1), 271–279.

167. Singh, A., et al. (1995). IL-6sR: Release from MCF-7 breast cancer cells and role in regulating peripheral oestrogen synthesis. *Journal of Endocrinology, 147*(2), R9–R12.

168. Macdiarmid, F., et al. (1994). Stimulation of aromatase activity in breast fibroblasts by tumor necrosis factor alpha. *Molecular and Cellular Endocrinology, 106*(1–2), 17–21.

169. Zhao, Y., et al. (1996). Tumor necrosis factor-alpha stimulates aromatase gene expression in human adipose stromal cells through use of an activating protein-1 binding site upstream of promoter 1.4. *Molecular Endocrinology, 10*(11), 1350–1357.

170. To, S. Q., et al. (2013). Involvement of early growth response factors in TNF-alpha-induced aromatase expression in breast adipose. *Breast Cancer Research and Treatment*.

171. To, S. Q., Knower, K. C., & Clyne, C. D. (2013). NF-kappaB and MAPK signalling pathways mediate TNF-alpha-induced early growth response gene transcription leading to aromatase expression. *Biochemical and Biophysical Research Communications, 433*(1), 96–101.

172. Deb, S., et al. (2004). Estrogen regulates expression of tumor necrosis factor receptors in breast adipose fibroblasts. *Journal of Clinical Endocrinology and Metabolism, 89*(8), 4018–4024.

173. Mendelson, C. R., et al. (1986). Growth factors suppress and phorbol esters potentiate the action of dibutyryl adenosine 3',5'-monophosphate to stimulate aromatase activity of human adipose stromal cells. *Endocrinology, 118*(3), 968–973.

174. Lueprasitsakul, P., Latour, D., & Longcope, C. (1990). Aromatase activity in human adipose tissue stromal cells: Effect of growth factors. *Steroids, 55*(12), 540–544.

175. Schmidt, M., & Loffler, G. (1994). Induction of aromatase in stromal vascular cells from human breast adipose tissue depends on cortisol and growth factors. *FEBS Letters, 341*(2–3), 177–181.

176. Su, B., et al. (2011). Growth factor signaling enhances aromatase activity of breast cancer cells via post-transcriptional mechanisms. *Journal of Steroid Biochemistry and Molecular Biology, 123*(3–5), 101–108.

177. Zhou, J., et al. (2001). Malignant breast epithelial cells stimulate aromatase expression via promoter II in human adipose fibroblasts: An epithelial-stromal interaction in breast tumors mediated by CCAAT/enhancer binding protein beta. *Cancer Research, 61*(5), 2328–2334.

178. Sparano, J. A., et al. (2012). Obesity at diagnosis is associated with inferior outcomes in hormone receptor-positive operable breast cancer. *Cancer, 118*(23), 5937–5946.

179. Chen, S., et al. (2012). Obesity or overweight is associated with worse pathological response to neoadjuvant chemotherapy among Chinese women with breast cancer. *PLoS ONE, 7*(7), e41380.

180. Sendur, M. A., et al. (2012). Efficacy of adjuvant aromatase inhibitor in hormone receptor-positive postmenopausal breast cancer patients according to the body mass index. *British Journal of Cancer, 107*(11), 1815–1819.

181. Ewertz, M., et al. (2012). Obesity and risk of recurrence or death after adjuvant endocrine therapy with letrozole or tamoxifen in the breast international group 1-98 trial. *Journal of Clinical Oncology, 30*(32), 3967–3975.

182. Suzuki, R., Saji, S., & Toi, M. (2012). Impact of body mass index on breast cancer in accordance with the life-stage of women. *Front Oncol, 2*, 123.

183. Pfeiler, G., et al. (2013). Efficacy of tamoxifen+/-aminoglutethimide in normal weight and overweight postmenopausal patients with hormone receptor-positive breast cancer: An analysis of 1509 patients of the ABCSG-06 trial. *British Journal of Cancer, 108*(7), 1408–1414.

184. Sestak, I., et al. (2010). Effect of body mass index on recurrences in tamoxifen and anastrozole treated women: An exploratory analysis from the ATAC trial. *Journal of Clinical Oncology, 28*(21), 3411–3415.

185. Wolters, R., et al. (2012). Endocrine therapy in obese patients with primary breast cancer: Another piece of evidence in an unfinished puzzle. *Breast Cancer Research and Treatment, 131*(3), 925–931.

186. Folkerd, E. J., et al. (2012). Suppression of plasma estrogen levels by letrozole and anastrozole is related to body mass index in patients with breast cancer. *Journal of Clinical Oncology, 30*(24), 2977–2980.

187. Diorio, C., et al. (2012). Aromatase inhibitors in obese breast cancer patients are not associated with increased plasma estradiol levels. *Breast Cancer Research and Treatment, 136*(2), 573–579.

188. Harvie, M., & Howell, A. (2006). Energy balance adiposity and breast cancer - energy restriction strategies for breast cancer prevention. *Obesity Reviews, 7*(1), 33–47.

189. Cleary, M. P., et al. (2002). Weight-cycling decreases incidence and increases latency of mammary tumors to a greater extent than does chronic caloric restriction in mouse mammary tumor virus-transforming growth factor-alpha female mice. *Cancer Epidemiology, Biomarkers & Prevention, 11*(9), 836–843.

190. Jones, S. B., et al. (2012). Effect of exercise on markers of inflammation in breast cancer survivors: The Yale exercise and survivorship study. *Cancer Prevention Research (Philadelphia)*.

191. Lof, M., Bergstrom, K., & Weiderpass, E. (2012). Physical activity and biomarkers in breast cancer survivors: A systematic review. *Maturitas, 73*(2), 134–142.

192. Jones, M. E., et al. (2013). Changes in estradiol and testosterone levels in postmenopausal women after changes in body mass index. *The Journal of Clinical Endocrinology & Metabolism*.

193. Esposito, K., et al. (2003). Effect of weight loss and lifestyle changes on vascular inflammatory markers in obese women: A randomized trial. *The Journal of the American Medical Association, 289*(14), 1799–1804.

194. Dandona, P., et al. (1998). Tumor necrosis factor-alpha in sera of obese patients: Fall with weight loss. *Journal of Clinical Endocrinology and Metabolism, 83*(8), 2907–2910.

195. Bastard, J. P., et al. (2000). Elevated levels of interleukin 6 are reduced in serum and subcutaneous adipose tissue of obese women after weight loss. *Journal of Clinical Endocrinology and Metabolism, 85*(9), 3338–3342.

196. Cancello, R., et al. (2005). Reduction of macrophage infiltration and chemoattractant gene expression changes in white adipose tissue of morbidly obese subjects after surgery-induced weight loss. *Diabetes, 54*(8), 2277–2286.

197. Ashrafian, H., et al. (2011). Metabolic surgery and cancer: Protective effects of bariatric procedures. *Cancer, 117*(9), 788–799.

198. Christou, N. V., et al. (2008). Bariatric surgery reduces cancer risk in morbidly obese patients. *Surgery for Obesity and Related Diseases, 4*(6), 691–695.

199. Bodmer, M., et al. (2010). Long-term metformin use is associated with decreased risk of breast cancer. *Diabetes Care, 33*(6), 1304–1308.

200. Beck, E., & Scheen, A. J. (2010). (Anti-cancer activity of metformin: New perspectives for an old drug). *Rev Med Suisse 6*(260), 1601–1607.

201. Niraula, S., et al. (2012). Metformin in early breast cancer: A prospective window of opportunity neoadjuvant study. *Breast Cancer Research and Treatment, 135*(3), 821–830.

202. Rena, G., Pearson, E. R., & Sakamoto, K. (2013). Molecular mechanism of action of metformin: Old or new insights? *Diabetologia, 56*, 1898–1906.

203. Buzzai, M., et al. (2007). Systemic treatment with the antidiabetic drug metformin selectively impairs p53-deficient tumor cell growth. *Cancer Research, 67*(14), 6745–6752.

204. Zhuang, Y., & Miskimins, W. K. (2008). Cell cycle arrest in Metformin treated breast cancer cells involves activation of AMPK, downregulation of cyclin D1, and requires p27Kip1 or p21Cip1. *Journal of Molecular Signaling, 3*, 18.

205. Ben Sahra, I., et al. (2010). Targeting cancer cell metabolism: The combination of metformin and 2-deoxyglucose induces p53-dependent apoptosis in prostate cancer cells. *Cancer Research, 70*(6), 2465–2475.

206. Phoenix, K. N., Vumbaca, F., & Claffey, K. P. (2009). Therapeutic metformin/AMPK activation promotes the angiogenic phenotype in the ERalpha negative MDA-MB-435 breast cancer model. *Breast Cancer Research and Treatment, 113*(1), 101–111.

207. Dowling, R. J., et al. (2007). Metformin inhibits mammalian target of rapamycin-dependent translation initiation in breast cancer cells. *Cancer Research, 67*(22), 10804–10812.

208. Brown, K. A., et al. (2010). Metformin inhibits aromatase expression in human breast adipose stromal cells via stimulation of AMP-activated protein kinase. *Breast Cancer Research and Treatment, 123*(2), 591–596.

209. Samarajeewa, N. U., et al. (2011). Promoter-specific effects of metformin on aromatase transcript expression. *Steroids, 76*(8), 768–771.

210. Rubin, G. L., et al. (2000). Peroxisome proliferator-activated receptor gamma ligands inhibit estrogen biosynthesis in human breast adipose tissue: Possible implications for breast cancer therapy. *Cancer Research, 60*(6), 1604–1608.

211. Rubin, G. L., et al. (2002). Ligands for the peroxisomal proliferator-activated receptor gamma and the retinoid X receptor inhibit aromatase cytochrome P450 (CYP19) expression mediated by promoter II in human breast adipose. *Endocrinology, 143*(8), 2863–2871.

212. Chand, A. L., et al. (2010). The orphan nuclear receptor LRH-1 promotes breast cancer motility and invasion. *Endocrine-Related Cancer, 17*(4), 965–975.

213. Thiruchelvam, P. T., et al. (2011). The liver receptor homolog-1 regulates estrogen receptor expression in breast cancer cells. *Breast Cancer Research and Treatment, 127*(2), 385–396.

214. Chand, A. L., et al. (2012). The orphan nuclear receptor LRH-1 and ERalpha activate GREB1 expression to induce breast cancer cell proliferation. *PLoS ONE, 7*(2), e31593.

215. Busby, S., et al. (2010). Discovery of Inverse Agonists for the Liver Receptor Homologue-1 (LRH1; NR5A2), in *Probe Reports from the NIH Molecular Libraries Program,* Bethesda, MD.

216. Purohit, A., et al. (1999). Inhibition of tumor necrosis factor alpha-stimulated aromatase activity by microtubule-stabilizing agents, paclitaxel and 2-methoxyestradiol. *Biochemical and Biophysical Research Communications, 261*(1), 214–217.

217. Deb, S., et al. (2006). A novel role of sodium butyrate in the regulation of cancer-associated aromatase promoters I.3 and II by disrupting a transcriptional complex in breast adipose fibroblasts. *Journal of Biological Chemistry, 281*(5), 2585–2597.

218. Alvarez-Garcia, V., et al. (2012). Melatonin interferes in the desmoplastic reaction in breast cancer by regulating cytokine production. *Journal of Pineal Research, 52*(3), 282–290.

219. Knower, K. C., et al. (2012). Melatonin suppresses aromatase expression and activity in breast cancer associated fibroblasts. *Breast Cancer Research and Treatment, 132*(2), 765–771.

220. Diaz-Cruz, E. S., Shapiro, C. L., & Brueggemeier, R. W. (2005). Cyclooxygenase inhibitors suppress aromatase expression and activity in breast cancer cells. *Journal of Clinical Endocrinology and Metabolism, 90*(5), 2563–2570.

221. Brueggemeier, R. W., et al. (2005). Translational studies on aromatase, cyclooxygenases, and enzyme inhibitors in breast cancer. *Journal of Steroid Biochemistry and Molecular Biology, 95*(1–5), 129–136.